LIMESTONE LEDGES AND SANDSTONE SPIRES IN DARK CANYON

A SWEEP OF THE COLORADO, BENEATH HATCH POINT MESA

EARLY SUMMER IN ARCHES NATIONAL MONUMENT

A COPSE OF GAMBEL OAK IN THE LA SAL FOOTHILLS

WEATHERED TOWERS IN WINTER, NEAR MOAB, UTAH

SNOW AT FOOT OF SETTING HEN BUTTE, MONUMENT VALLEY

MOONSET OVER THE SAN JUAN CLIFFS IN JOHNS CANYON

ISLAND IN THE SKY MESA, BACKED BY THE LA SAL MOUNTAINS

THE ART OF SEWING
THE OLD WEST
THE EMERGENCE OF MAN
THE AMERICAN WILDERNESS
THE TIME-LIFE ENCYCLOPEDIA OF GARDENING
LIFE LIBRARY OF PHOTOGRAPHY
THIS FABULOUS CENTURY
FOODS OF THE WORLD
TIME-LIFE LIBRARY OF AMERICA
TIME-LIFE LIBRARY OF ART
GREAT AGES OF MAN
LIFE SCIENCE LIBRARY
THE LIFE HISTORY OF THE UNITED STATES
TIME READING PROGRAM
LIFE NATURE LIBRARY
LIFE WORLD LIBRARY
FAMILY LIBRARY:
  THE TIME-LIFE BOOK OF THE FAMILY CAR
  THE TIME-LIFE FAMILY LEGAL GUIDE
  THE TIME-LIFE BOOK OF FAMILY FINANCE

CANYONS AND MESAS

THE AMERICAN WILDERNESS/TIME-LIFE BOOKS/NEW YORK

BY JEROME DOOLITTLE
AND THE EDITORS OF TIME-LIFE BOOKS

WITH PHOTOGRAPHS BY WOLF VON DEM BUSSCHE

THE AMERICAN WILDERNESS

SERIES EDITOR: Charles Osborne
Editorial Staff for *Canyons and Mesas:*
Text Editor: Jay Brennan
Picture Editor: Kaye Neil
Designer: Charles Mikolaycak
Staff Writers: Simone D. Gossner,
Don Nelson, Michele Wood
Chief Researcher: Martha T. Goolrick
Researchers: Doris Coffin, John Hamlin,
Villette Harris, Beatrice Hsia,
Suzanne Wittebort, Editha Yango
Design Assistant: Vincent Lewis

Editorial Production
Production Editor: Douglas B. Graham
Assistant: Gennaro C. Esposito
Quality Director: Robert L. Young
Assistant: James J. Cox
Copy Staff: Rosalind Stubenberg (chief),
Barbara Quarmby, Mary Ellen Slate,
Florence Keith
Picture Department: Dolores A. Littles,
Joan Lynch
Traffic: Feliciano Madrid

Valuable assistance was given by the
following departments and individuals of
Time Inc.: Editorial Production, Norman Airey;
Library, Benjamin Lightman; Picture
Collection, Doris O'Neil; Photographic
Laboratory, George Karas; TIME-LIFE News
Service, Murray J. Gart.

The *Author:* Jerome Doolittle first became interested in deserts and their flora and fauna while covering irrigation projects in the Sahara for the United States Information Agency in the 1960s. Before that, Mr. Doolittle worked as a reporter, columnist and assistant city editor of *The Washington Post.* As a freelance writer, he has contributed to a number of national publications, including *Esquire* and *Holiday.* He lives in Connecticut.

The *Photographer:* Wolf von dem Bussche, a native of Germany who is now an American citizen, studied art history at Columbia College in New York and after several years of painting became a professional photographer in 1968. Since then, he has had four one-man shows and been a participant in several group shows in principal U.S. cities. He contributed major picture essays to the LIFE Library of Photography. For The American Wilderness series, he has photographed pictorial essays that were published in *Cactus Country, The Okefenokee Swamp* and *The Ozarks.*

The *Cover:* The glowing bulk of Merrimac Mesa rears above a winter morning's mist in the canyon country of Utah. Weathering and the gradual effects of gravity cause the bizarre conformation of the rocks on the cliff in the foreground and of the mesa-top sandstones, which are called voodoos for their grotesque shapes.

Contents

1/ The Canyon Lands   20

A Successful Struggle for Life   34

2/ The Rocks   46

A Natural Architecture   62

3/ The Forbidding Country   76

A Nature Walk down Johns Canyon   92

4/ The Mesas   104

5/ The Rivers   120

The Power of a River   140

6/ The Sacred Mountain   152

Spirits of the Rainbow Bridge Trail   166

Bibliography   180

Acknowledgments and Credits   181

Index   182

A Stony, Arid Land of Splendors

Seamed by arroyos and canyons, studded with the huge forms of mesas, buttes and towering pinnacles, the desolate but magnificent region described in this volume (shaded in the map above) occupies approximately 32,000 square miles at the center of the Colorado Plateau. Its deserts and mountains are drained by the Colorado River and its tributaries. Though broken here and there by extensive forests of pine and juniper, the Canyon Lands are comparatively bare of vegetation. The tops of mesas and plateaus are as much as a mile above sea level, where they receive only an average 10-inch rainfall, and where they are exposed to seasonal extremes of heat and cold.

Of the rivers (shown in blue), only the Colorado, the Green and the San Juan usually flow year round. Lesser streams tend to dry up completely during the dry season.

The few roads that cross the canyon country are shown in white; red lines enclose such federally protected areas as national parks, forests, monuments and recreation areas, as well as numerous Indian reservations.

1/ The Canyon Lands

The finest workers in stone are not copper or steel tools, but the gentle touches of air and water working at their leisure with a liberal allowance of time.

HENRY DAVID THOREAU/ *A WEEK ON THE CONCORD AND MERRIMACK RIVERS*

Look at a map, and all roads will be seen to approach the canyon and mesa country of southeastern Utah but few to enter it.

This is the heart of a tough, high, lonely and lovely land—the 150,000 square miles of the Colorado Plateau. An empty wilderness nearly the size of California, the plateau surrounds the point where the states of Colorado, Utah, New Mexico and Arizona meet. Only around the edges of the plateau does man live in any numbers, and even there his presence is not particularly obtrusive—the most populous town is Flagstaff, Arizona, with 26,117 inhabitants. Other major urban centers are Farmington, New Mexico; Grand Junction, Colorado; Gallup, New Mexico; and Durango, Colorado. Outside of these relatively small towns, most of the plateau is uninhabited, and visited only by an occasional Navajo sheepherder. Many areas, it is safe to say, have never been visited at all, even by Indians.

This is particularly true of the region physiographers call the Canyon Lands, a region roughly bounded by the San Juan Mountains on the east, the Book Cliffs on the north, the Aquarius Plateau on the west, and the San Juan and Colorado rivers on the south. The eastern part of the section lies in southwestern Colorado and its western part lies mainly in southeastern Utah, with a small finger sticking south into Arizona along the Colorado River.

"Perhaps no portion of the earth's surface is more irremediably sterile," Professor J. S. Newberry of Columbia College wrote in 1860, "none more hopelessly lost to human occupation." Nothing has happened in the past century or so to prove the professor seriously wrong. The heart of the canyon and mesa country remains unchanged—sterile, harsh, empty and difficult of passage.

And magnificent.

Professor Newberry saw this, too. A little later in his account of the 1859 expedition to the region on which he served as geologist, the professor wrote, " ... the surface was diversified by columns, spires, castles, and battlemented towers of colossal but often beautiful proportions, closely resembling elaborate structures of art, but in effect far surpassing the most imposing monuments of human skill."

The professor also realized that the country was as perilous and demanding as it was grand. "After reaching the elevated point from which we obtained this view," he wrote, "I neglected to take the rest I so much needed. . . . Standing on the highest point, I made a panoramic sketch of the entire landscape. The effort had, however, nearly cost me dear; for before I had completed the circle of the horizon I was seized with dreadful headache, giddiness, and nausea, and, alone as I then was, had the greatest difficulty in rejoining my companions."

The professor discerned clearly the three chief elements of the country: its barrenness, its vast and desolate beauty—and its danger. To those who love the canyon and mesa country, all three elements are linked and indivisible. Without any one of them, the love could not exist, or not quite in the same form. "Why do you do it?" a stranger to the canyons once asked a friend of mine who was about to set out on a hard hike over terrain he did not know. "Why would you want to risk your life going someplace nobody has ever been to before?"

My friend did not even try to answer the question, but I have tried, in my mind, ever since. Perhaps my friend was going because the trip would be hard and risky, because the route was unknown, and because very few people, certainly, had ever made that trip before. And because, if you love the canyon country, you not only accept such things, you require them. Is that an answer? I don't know. But I do know that if the trials and the risks were not there, the Canyon Lands would not be the same thing to my friend—or to me. It would be like Cleveland, or Yellowstone with its garbage-eating bears.

I once heard about a man who went on a relatively undemanding backpacking trip with three hikers who know the country well. The man, new to the canyons and mesas, came apart under the brutal, indifferent weight of the land. Hot, thirsty, frightened, exhausted, he reverted to the spoiled little boy he must once have been. He complained about things that it does no good to complain about and he quarreled with the three men who were trying to nurse him along safely. "Jesus," he said to them at one point, "you guys must have been in the Marines." If I understand him right, what he meant by a Marine was one who seeks certification of his manhood by joining a tough gang in which he will be trained to accept hardship and display courage as a member of a team.

The three men with him had all moved to the Colorado Plateau from other parts of the country, to be near the canyons they loved. Whenever they could, they went out to try themselves against those canyons. With a hint of self-mockery, they called themselves and others like them canyon runners or canyon rats. None of the three had ever been in the Marine Corps.

Typically, the men who go out on foot in that country are loners rather than team players. If they find any toughening of mind or body to be necessary they will do the job by themselves, with no need of a drill instructor. If they want to be brave, they require a lonely place to do it in. A canyon rat's only audience is himself, tougher than the Corps, but more satisfying. One young ranger at Canyonlands National Park, learning that I was working on this book, said, "Make the place sound awful. Too goddam many people here as it is."

I don't think I can make it sound awful, though, because for me it is not, any more than it is for the ranger. But I will not limit myself to the beauties of the canyon country, either. That would be like ignoring scars and tattoos that everybody can see on your lady friend, in favor of describing only her very fine bone structure. And fine it no doubt is, though she is full of heat that sucks the water out of the blood, rocks that turn under the feet, snakes to bite, thorns to tear, flash floods to drown a man, and blizzards to freeze him. Another friend of mine, normally no more emotionally demonstrative than any other American male, once wrote to me, "These lonely places are my mistress, and I love her deeply. . . ."

I can see why he feels that way.

But does any of this answer the question? If it does, I would be very much surprised. Maybe the truth is that anybody who has to ask the question will not understand the answer.

In any event, let us begin with the tattoos and scars; we will get to the fine facial planes later.

Travel through the Canyon Lands is difficult—where it is not entirely impossible. More or less permanent streams run through some canyons, but you usually cannot get to the water. In one 135-mile stretch of the San Juan River, for example, livestock can reach the water at only three spots. The two principal crossings of the Colorado River, at Lee's Ferry and at Dandy Crossing, are 170 miles apart. The farming is marginal, the soil ranging from poor to nonexistent. The mineral resources are unimpressive, the pasturage sparse; water is in short supply, except when it comes along in a flash flood to wash away a house or a section of farmland.

The weather is unspeakable. The 144° annual temperature range —from 115°F. to -29°F.—recorded for San Juan County, Utah, is one of the widest in the world. Daily variations of 50° are not at all uncommon. Once, during the course of a single hour in March, I was snowed on, rained on, sleeted on, buffeted by high winds and broiled by the sun.

For lack of water and fertile soil, plants are spaced so widely apart that the land, seen from a distance, can look entirely naked. It can even look that way close up. In 1876, T. S. Brandegee of the U.S. Geological Survey wrote: "There are large areas with absolutely nothing growing upon them, and often, even along the streams, our day's journey would be lengthened four or five miles before grass could be found sufficient for a camping place." Even the cacti, Mr. Brandegee remarked, were small and not particularly abundant.

Nor is the animal population impressive; I have walked whole days without starting any ground dwellers larger than lizards. A book of mine has a series of maps giving the distribution of Western reptiles. The shaded areas, representing various species, approach the canyon and mesa country but, like the roads, seldom enter it. When I asked a zoologist about this, he agreed that this indeed seemed to be the case. Of the 70-odd species of snakes inhabiting Western North America, he could think of only a half dozen or so to be found in the area.

"On the other hand," he said, "remember who makes those habitat maps. The maps may just reflect the fact that the herpetologist population in there is pretty skimpy, too."

In the Wilderness Act of 1964, wilderness is established as "an area where the earth and its community of life are untrammeled by man, where man himself is a visitor who does not remain." By these standards, most of the Canyon Lands are indeed wilderness. They remain in that fortunate state, untrammeled by man even if trammeled

The torn, scoured face of the Colorado's canyon rises from the riverbank, providing a massive tableau of the land's history. Weathering and erosion carved the arches and bays that stand out against the cliffside, while the river itself created the horizontal terraces in cutting downward to successive bed levels over the ages.

pretty badly in spots by his sheep and cows, because nobody has yet figured a way to make much of a buck out of the area. The same factors that make the country so spectacularly beautiful also make it inhospitable to man and his projects: the whole plateau is high and dry. Most of it is a mile above sea level and a few mountains rise more than twice that high. Rainfall is very slight—well below 10 inches a year over much of the area, although the occasional mountain receives somewhat more. And when the rains do come, they are seldom gentle, cooling sprinkles. More often they are violent thunderstorms whose water strips away sand and soil while giving little more than a momentary and superficial wetting to the pavement rock beneath.

Moreover, when rain falls on high country it has a mile to descend before it returns to sea level. Thus its energy, both to cut into the rock and to carry away material already loosened by erosion and weathering, is far greater than that of rain falling on some coastal plain. Since the extreme aridity prevents formation of a protective skin of vegetation, the land lies naked to erosion. And erosion has done its worst, tearing away at the earth to an extent unknown anywhere else in the United States. No one knows how great a thickness of rock has been removed from the Colorado Plateau, but the guesses of geologists range up to 10,000 feet—a blanket of stone nearly two miles high. According to one estimate, erosion in the canyon country continues to lower its general elevation by about an inch every 440 years.

What remains is a place where the scenery, by and large, is upside down. Although here and there in the Canyon Lands, ranges of mountains stick up from the level of the plateau, much more common are the canyons that stick down. East and south of the Abajo Mountains in Utah lies what seems to be a featureless plain, punctuated only by the Carrizo Mountains of Arizona and the Ute Mountains of Colorado. But this expanse, called the Great Sage Plain, is not at all unbroken. Invisible from the mountaintops are some 20 canyons that cut deeply into the surface of Great Sage.

This country has been laid bare to the bones—a ruined battlefield that has been torn and ripped by nature much more effectively than man can yet manage. But life survives, and sometimes even thrives. Perhaps, after all, this is the secret of why those who love the canyon country do so in such immoderate fashion. The flower blooming where no flower should be able to grow; the toad, improbably alive miles from any permanent water; the kangaroo rat, which manages its life so clev-

STAGECOACH, 1939

Rattling through the barren majesty of Monument Valley, a stagecoach travels with cavalry in director John Ford's classic Stagecoach (left), the first movie made in the canyon country. Ford's lead was quickly followed by other directors who took full advantage of buttes and mesas as backdrops for scores of Western epics. Then, director George Stevens used the area to re-create Biblical Palestine for The Greatest Story Ever Told and sent actors trudging over a makeshift bridge dressed up as ancient Israelites.

THE GREATEST STORY EVER TOLD, 1965

erly that it can get along for months without water. And my friend out there, too, not nearly so well adapted for survival as the others, but nevertheless still alive. Who can say that he hasn't earned the right to make a fool of himself?

Earning is a big part of the whole thing in the minds of most men who truly—and properly—love the region. Anyone can be impressed by the Canyon Lands, or admire them esthetically, or be awed by them. All this costs is the price of a movie ticket; the wild, savage scenery of Utah and Arizona has been the setting for a hundred cowboy pictures since 1938, when John Ford used Monument Valley as a backdrop for *Stagecoach.* And so, that way, the canyon and mesa country can be had cheap. But that way it becomes nothing more than an elaborate set, John Wayne at stage right, posse at stage left, coming up fast—all too carefully composed and theatrical to seem real.

The best way to learn to love the country is to search out, on foot, the land's smallnesses. From the rim, a canyon floor may indeed look sere, but once you have made your way down the wall, the bottomland may look green and fertile. Compared to an August meadow in New England, the most fertile canyon floor would still seem a desert, but these things are relative. The point is that next to the New England meadow lies another just as green, and next to that another meadow, and another. The green, secret corners of the Canyon Lands, though, are separated from one another by miles of barren walking. Each step drives up their value.

One spring day I was camped alone where Cross Canyon runs into the Colorado River. I was lying under my poncho, from which I had rigged a sort of flat-topped tent about two feet off the ground. Rain tapped on the poncho. I had got wet rigging the thing. The fire outside was getting wet, too, which made it smoke. The wind carried the smoke into my makeshift tent. Just enough light remained in the afternoon so that, by straining my eyes, I could read a paperback book: John Gardner's *Grendel,* the Beowulf epic told from the monster's point of view.

Next morning I made hot chocolate with water from the river, which was nearly the right color even before I put in the chocolate. The water contained so much silt that particles gritted between my teeth. My gear was still damp. I set my new $42 boots by the fire to dry out a little, not noticing that I had set them too close—until one started to smoke. By that time, melted black stuff was oozing out of its heel.

When I started up the Colorado River after breakfast, the day was so

dull that the morning sun cast no shadows. For once in this dry country, the air was heavy and muggy. The east bank of the river at that
point, and as far upstream as I could see, consisted of steep slopes of
boulders running right into the brown water. Better to climb cliffs or
wade streams or go through quicksand than to walk on talus slopes like
these. The boulders and rocks are of uneven size so that the footing is
not level. Often they roll and wobble when you step on them and so
you constantly risk turning an ankle or breaking a leg—serious or even
fatal matters when you are alone.

The muggy heat made me take off my shirt; there was a white rime
of dried sweat on my pack straps. Since nothing grew on my endless
rock pile, it offered no shade. Once I saw a lizard, but that was all.
Only I was mad enough to be abroad on this sterile riverbank in this
heavy heat, pushing idiotically upstream along one of the world's major flows of liquid mud.

To tell the truth, I wasn't entirely pleased with the way things had
been going. But then, suddenly, I came across the rattlesnake, and everything was all right again.

It takes all kinds—as is well known; and I am the kind that is interested in, and seeks out, snakes. I had seen cobras and kraits and Rus-
sell's vipers when I was living in Southeast Asia during the '60s. I was
once stupid enough to get myself bitten by a poisonous tree viper in Ma-
laysia. But though I presently live in one of the few parts of Connecticut
where timber rattlers are still found, I had never before seen a rattlesnake in the wild.

I knew the sound, of course, and spotted the snake right away: a
midget faded rattlesnake, not two feet long, the color of brick dust. In
the shade of a ledge under a large boulder, the snake was forming a series of loops that traveled along its body the way they will along a rope
when you shake one end. The performance was full of menace, unless
you happened to notice that each of the angry looping movements carried the animal a few inches back under the boulder. In a moment the
snake had disappeared entirely from view, although the dry buzzing of
the rattle still hung in the air like the grin of the Cheshire cat. The noise
continued to sound from underground for several minutes more, until
the little snake got over its fright at seeing a monster.

I grant that seeing a midget faded rattlesnake in a rock pile may not
be everybody's idea of a magic moment, but it was mine. I felt greatly
improved. As I resumed my progress, the heat seemed less oppressive;
the water of the river turned from the previous vile shade to a healthy,

earthy color. Somewhat more objectively, the talus slope soon ended and walking became much easier.

Around the next bend in the Colorado I saw that I was coming up to Spanish Bottom, where Lower Red Lake Canyon meets the Colorado River. Here the canyon of the Colorado widens for a bit, so that there is a small, wooded plain on either side. I found a green bank shaded by cottonwoods and set down my pack.

No matter how often I do it in the course of a day, each time I take off my pack is like getting out of the Army all over again. I had put the pack down in a patch of lamb's-quarters, a wild green, good raw or cooked, and so I plucked a handful to eat while I went off scouting.

Bees murmured in the air. The plants underfoot gave off their scents as I crushed them in passing, and the perfumes of the different plants succeeded each other. In easier climates plants are so crowded together that a field smells like a field, the sum of its parts. But in the Canyon Lands the scents seem to separate out.

Soon I came to a channel cut by the stream that goes down Lower Red Lake Canyon—sometimes. Right now the stream was dried up, but the Colorado River itself was so high that its waters had backed partway up the stream bed. The channel had steep, sandy, collapsible walls higher than my head. The only way to the bottom would be to half-slide, half-fall; I went back to get my pack so as not to have to cross the channel again. Then I traversed above the point where the river water had reached. A clearing surrounded by cottonwoods sat on top of the little bluff on the other side, a good place to make camp. I took off my clothes there and hung them from a cottonwood snag.

When I slid back down the bank, the sand was soft and smooth against my skin, almost powdery. I waded downstream until the water wrapped me chest deep in coolness. Minnows, invisible in the brown river water, tickled as they tasted me. When I stepped, the mud squeezed up between my toes; when I stayed still, my feet settled comfortably down into it, ankle deep.

Once the coolness of the water had begun to turn to chilliness I got out. I lay down in the sand on top of the bluff to let myself be dried off by the mild breath of a breeze. I felt the air as neither hot nor cold; was I its temperature, or was it mine? Beside me was a drift of dead cottonwood leaves, and a continuous low rustling came from it. When I brushed a few leaves aside, I saw that the noise came from crickets, dozens of them, rummaging around for food.

Below a stand of tamarisk, the silt-laden Colorado eases past a sandbar on the shore of Spanish Bottom near Lower Red Lake Canyon.

Another noise came from above me, the surprisingly loud whir of a hummingbird. I fumbled my field glasses out of the pack beside me, and made the hummingbird larger. But the tiny bird was silhouetted black against the sky, so that I could not tell which of the several species found in the canyon country it was. People who know more about these things can deduce the breeding presence of hummingbirds from the occurrence of certain flowers. Often these flowers are red, like Indian paintbrush and penstemon.

Hummingbirds are said to perceive red as a particularly vivid color; by giving off such a brilliant signal a red flower increases its chances of being pollinated by hummingbirds in search of nectar. There were certainly no red flowers in the top branches of the cottonwood, nor any flowers at all as far as I could see. But the hummingbird had some lengthy business up there nonetheless. Perhaps it was looking me over. It would hover in one spot for a little bit and then dart away to a new place to hover some more. At last it made a definitive, final dart and disappeared from my world, the glade.

Now that the binoculars were out, I propped my head against the log behind me and began to sweep them around the clearing. A moving bit of yellow across the glade caught my attention. A bumblebee with a brilliant solid-yellow back was at work on some daisies. Bees apparently not only do not see red as a particularly vivid color; they see no red at all. But they do discern shades of yellow, though not as yellow, but as colors unimaginable to us. And they sense ultraviolet rays (which our naked eye cannot detect) with great vividness. Thus, to a bee, a daisy must look utterly unlike our experience of the thing. Some desert daisies, whose petals look uniformly yellow to us, actually absorb ultraviolet rays except at the tips, where spots reflect it. The bee therefore presumably sees the daisy as a halo of ultraviolet dots surrounding the food source at the center of the flower.

But one need not have an entomologist's grasp of these things in order to watch a bumblebee. As I looked at this bee and these daisies, I had no idea of how a daisy looks to a bee, though I have since happened across the above curiosities in a book. The fact is I have always been regrettably inept at imagining unimaginable colors. For the moment, though, I was lying naked in the sun, ignorant but perfectly happy. Optics concerned me only insofar as I got to thinking about how looking through binoculars puts things into a sort of frame and makes pictures out of them by excluding the rest of the world. And so I began making pictures for myself all over the glen by looking here and there through my toy. Fifteen feet away was a mushroom, which I composed carefully against the sand. Normally the size of a saucer, the mushroom cap became even bigger in my picture. The fungus had risen up out of the ground so recently that its top was still covered with sand. I turned to frame an Audubon's warbler and a spiny lizard, both patrolling different parts of the same cottonwood tree across the way. Next I examined a stand of tall reeds on the other side of the glade and found a Chinese scroll painting. The reeds formed a stylized, two-dimensional tracery that looked just like Sung Dynasty paintings of bamboo leaves. A blister on my toe, a very fine one, was too close to get into focus.

Finally I put aside the field glasses and, for a long time, just looked at the pale blue of the sky, and felt the fine sand warm under me and listened to the crickets in the dry cottonwood leaves.

Cool water, hummingbirds, bees, pictures in binocular scale—these things could exist for anyone in Spanish Bottom. But the rain the night before, the smoke, the grit in my mouth, the heavy heat, the endless rock piles had also created Spanish Bottom for me. The place exists, all right, pretty much the way I tell about it, and you can get to it by drifting down the Colorado on a raft or even by flying there in a helicopter. You can get to other places that are a lot like Spanish Bottom by Jeep or motorcycle or speedboat—or by station wagon, for all I know. And there you may see the same things I did.

But you will not have been in the same place.

A Successful Struggle for Life

Mile after mile of sunbaked rock and parched sand confront the traveler in the Canyon Lands wilderness in southeastern Utah. Indeed, much of the country resembles high desert, waterless and apparently devoid of plant life. But close inspection reveals that there are actually very few areas where plants have been totally unable to gain and maintain at least a foothold.

Even in the most hostile places, tough, drought-resistant flora send roots down and out into the shallow soils and squeeze life from the few drops of underground water. Moreover, the plants are always ready to exploit the occasional fall of rain or the ground runoff from a flash flood.

In June, the area's plant life is at its height. Having made full use of snowmelt and spring rains to spur growth before the worst of summer's desiccating heat arrives, most plants are dressed in full greenery and adorned with flowers. But even in June, local differences in the amount of available water—proximity to a stream, or a rise in altitude, which provides greater rainfall—make for dramatic contrasts from place to place in the forms and appearances of the various plants.

Between the Colorado River and the La Sal Mountains, contrasting landscapes—depicted on the following pages—contain plants whose lives and looks are finely tuned to their environments.

Squawbush, mountain mahogany and hymenoxys, pictured at right in a typically hostile canyon-country habitat, maintain their lives with a neat combination of adaptations. All three conserve moisture by keeping low profiles and by producing small leaves with leathery, hairy or waxy surfaces that hold evaporation to a minimum. In addition, such plants usually are spaced far apart so their root systems do not poach on one another in their use of the sparse supply of water.

In the valleys (page 39) and on the banks of intervening streams (page 40), where water is available in greater quantity, the struggle for survival is less arduous. In these places plants are able to grow more abundant leaves and live in much closer proximity. But the richest growing ground of all is on the north slope of the La Sal foothills (pages 44-45). There ground plants and trees, luxuriating in deeper, more nutritious soil and rainfall of 20 or more inches a year, largely escape the exigent conditions that prevail nearby.

Three hardy plant species survive in their desert niches beneath a butte near Dead Horse Point, Utah. In the foreground a mountain-mahogany shrub and a hymenoxys plant topped by yellow blossoms flourish in a common bed. At left a squawbush, sometimes called skunkbush because of the foul odor of its leaves, grows on a rocky patch of ground. All three share the meager runoff of water from snow and rain that fall on the butte.

AN EVENING-PRIMROSE IN PARCHED SOIL

A TWIN-POD IN RED SAND

Denizens of the Mesas and Valleys

For lack of rain, mesa tops and valleys are among the driest places in the Canyon Lands. Of the two habitats, the valleys are often more congenial since they are likely to trap eroded soil and retain moisture. Worst of all are those mesas covered only with bare rock or clay, for they are brutally exposed to sun and wind, and hold virtually no water.

Nevertheless, even on such mesa tops a surprising number of plants manage to survive, and yucca, piñon pine and juniper may actually thrive —though the grizzled limbs of the dead juniper shown at right bear witness to the uncertainty of life. These plants cope with aridity by deploying the most intricate root systems, which probe down deeply or spread laterally as they absorb water. Juniper roots may sometimes reach out as far as 50 feet from the tree. To hoard whatever moisture it can absorb, the evening-primrose (*top left*) opens its fragrant blossoms only at night, while the twin-pod (*bottom left*) has leaves with a downy coating that reduces evaporation.

Denser growth in such lowlands as Castle Valley (*overleaf*) confirms the existence of more hospitable conditions. Here the soil is fertile enough to support a variety of plants—including hawkweed, wild geranium, scurfpea, sow-thistle and bindweed. These weeds' close-knit roots bind the soil and allow successive generations to enrich their own beds with the mulch of dead leaves and stems.

Harriman yucca plants raise stalks, festooned with blossoms, that contrast with a mesa top's sturdy piñon pine and sprawled dead juniper.

HAWKWEED

WILD GERANIUM

SOW-THISTLE

BINDWEED

Castle Rock towers over its valley, fringed with a thicket of juniper and blooming with a ground cover of scurfpea (foreground).

Near a dead Fremont poplar (left), blackened skeletons of burned-over tamarisks mingle with new growth thrusting from unburned roots

Interlopers on the Banks of the Colorado

Where its canyon walls widen, the verges of the Colorado are lush with greenery. Cheat grass, fleabane and penny cress crowd the banks. With them flourishes a small tree, the tamarisk, a European import that has spread through much of the West.

Yet the riverbank area receives no more rainfall than Castle Valley. The plants' success along the Colorado lies in the fact that, in many places, the water table lies only two or three feet below the surface. Constantly replenished by seepage from the river, this supply is easily reached by the roots of the larger plants, which shade the smaller ones from the sun. Despite these advantages, the riverbank plants are by no means fragile. The cheat grass and penny cress have taken over from such species as perennial wheat grasses, which cattle have grazed beyond the plants' ability to recover. The tamarisk has supplanted such trees as the Fremont poplar, cut for firewood and fenceposts. These newcomers to the riverbanks utilize water efficiently; the vigorous grasses can withstand grazing, and the tamarisk survives because it is useless to man.

The present plants can even endure fires set by lightning—or by men. Though the tamarisk at left has been burned aboveground, fire could not reach the roots, which have put forth fresh foliage. The grasses, too, will probably revive, and burned plants will decompose, enriching the soil and encouraging new growth.

FLEABANE

PENNY CRESS

Sagebrush, Indian paintbrush in bloom and a young white fir dress the gently sloping foothills of the snow-capped La Sal Mountains.

The Fortunate Foothills

The foothills of the La Sal Mountains owe their great abundance of such comparatively moisture-demanding plants as phlox, Indian paintbrush and balsamroot to a combination of fortunate circumstances. The hills lie to the west of the main mountains, over which the winds rise, cool off and dump their moisture, sprinkling the hillsides with nearly three times as much rainfall as some nearby areas receive. In addition, the elevation of the hills gives them a temperature that averages 40° cooler than the mesa tops. Thus, plants that have established themselves on the hills are far less likely to lose moisture through evaporation. Finally, the foothills are carpeted deep in fertile soil that is replenished by sediment washed down from the slopes of the parent mountains.

Since the sun strikes obliquely on the north-facing slopes (*pages 44-45*), summer temperatures there are even cooler than on the southern slopes. Here are plants that can only thrive in cool, damp places: elderberry, golden pea and the graceful aspen. But the most dramatic evidence of the benign fertility of these north slopes is revealed in what is virtually the area's only true pattern of forest succession. Springing up in places beneath the aspens are the seedlings of Douglas fir, white fir and spruce; eventually, if conditions remain favorable, these may supplant the aspens, which cannot grow in the conifers' dense shade.

BLADDERPOD

BALSAMROOT

PINK LONGLEAF PHLOX

WHITE LONGLEAF PHLOX

ELDERBERRY

GOLDEN PEA

On the northern slopes of the La Sal foothills, quaking aspen trees shade a mixture of snowberry bushes, bluegrass and dandelion.

2/ The Rocks

*When we think of rock we usually think
of stones, broken rock, buried under soil and plant life, but
here all is exposed and naked.*   EDWARD ABBEY/ *DESERT SOLITAIRE*

All day we had been driving through falling snow. The snow made the desert into a stranger, a thing of soft mounds instead of angular bones. It blurred the sharp lines and hid the scorched colors. But the snow turned to rain as we drew near the Colorado River, so that the pale red of the land reappeared. By the time we reached Hite Marina on Lake Powell, our departure point for a five-day trip into the wild country to the north, the rocks were naked again, glistening cold and wet.

With the snow gone, we could see millions of years of the earth's history standing, exposed, on the other side of the lake. The cliff rising out of the water opposite us was typical of the formations that bring geologists from all over the world to study the Canyon Lands. At the top of the cliff lay the youngest strata of rock and at the bottom the oldest, all neatly arranged in descending order. Such orderly progression is sometimes called wedding-cake geology because of its neat layering. Here, to an extent unmatched elsewhere, the layers of rock are clearly visible; here the rivers have carved so deeply into the earth that they have brought to view its evolution over hundreds of millions of years.

The greatest canyon of all is the one cut by the Colorado River—so extensive that different stretches bear different names: along the lower river is the Grand Canyon; next upstream comes Marble Canyon; then Glen Canyon (most of it now submerged under Lake Powell); and up north of Hite Marina looms Cataract Canyon. The oldest rocks laid bare by the Colorado are the Vishnu schists in Grand Canyon, remnants of a huge mountain chain formed nearly two billion years ago. The most ancient in Cataract Canyon are the outcroppings that have given their name to the tributary canyon for which we were headed —Gypsum Canyon. Dating from the Middle Pennsylvanian Period, when the region was covered by a shallow sea, these have an age of about 300 million years. To reach Gypsum Canyon we had to travel by motorboat 28 miles up Lake Powell and the Colorado River itself. From the canyon mouth, we planned to take a 35-mile hike, through ravines and over mesas, ending at a spot on Dark Canyon Plateau where we had left a truck. With me were Haskell Rosebrough, a lawyer, Frank Nordstrom, a pediatrician, and Jim Fassett, a government geologist.

Silt had turned the upper waters of Lake Powell into a thin mud. Bursts of wind-driven rain splattered into the brown fluid as our boat knifed along between the black-streaked cliffs of this section of Cataract Canyon. The streaks, known as desert varnish, make parts of the canyon country look as though they have been dipped into printer's ink and then hung out to drip dry. The discolorations are made up largely of various minerals brought to the surface by seepage from underground water. This varnishing is fairly new. In some places, the oxides have been deposited on top of pictures chipped into canyon walls by the Anasazi Indians, who lived here as recently as the 13th Century.

Where there was no varnish, the walls of Cataract Canyon were the red color of Halgaito shale, which forms some of the most spectacular scenery of the Canyon Lands. Some 280 million years ago, the area was a coastal plain roamed by four-legged reptiles, the forebears of the dinosaurs. To the west was the ocean; to the northeast were huge mountains, called the Ancestral Rockies, that rose and were worn away before the creation of the present Rocky Mountains. The torrents flowing westward from the Ancestral Rockies were rapid enough to move pebbles, cobbles and even boulders. But when the water hit the flat coastal plains, it slowed and the heavier material dropped out. Only fine sand and mud were left by the time the streams reached the region of the present-day Colorado River. These particles were rich in iron, which oxidized in the air to make the rusty hues now before us.

As we moved upstream, we saw the red rock giving way to laminations of sandstone that dipped sharply in a generally southeast direction. These are what the geologists call cross-bedding, exposed where erosion has cut through ancient coastal sand dunes or offshore sandbars. The lamination lines represent the demarcation between the

thin layers of sand that make up the dune. Because of the angle of the cross-bedding, geologists feel that the sands forming these rocks were carried to their present location from the northwest—either by wind or by ocean currents. They now lie as a sort of pavement on the mesa tops, the softer rocks having been largely eroded away. The strata making up the pavement are called, collectively, Cedar Mesa sandstone.

Our planned hike would take us through something like 30 million years of geological time—all of it to be read in the canyon walls, from the rocks under our feet, and in the numerous fossils that were embedded in the rock. The boat let us off at the point where the Colorado River has cut down to the ancient gypsum, which, from a distance, gleams white in the sunshine. For Rosebrough, Nordstrom and me, this meant only that for a mile or two up Gypsum Canyon the dissolved minerals in its stream would make the water so laxative as to be undrinkable, but to geologist Jim Fassett, these outcroppings held a particular interest: they are the trace of subterranean layers of salt and gypsum that are a major chapter in the complex geological history of the canyon country. Here in Gypsum Canyon we were at the western edge of a formation of salt and gypsum that covers a little more than 11,000 square miles of southwestern Colorado and southeastern Utah.

During Pennsylvanian time, when that shallow sea still covered the canyon country, intense evaporation made the waters (like those of Great Salt Lake or the Dead Sea) much more salty than normal sea water. Halite (common salt) and gypsum both precipitate out of water when the solution reaches a certain degree of concentration. Since the precipitation point is nearly the same for both, they were laid down in alternating layers on the sea bottom until they formed beds more than a mile thick in places. Somewhat later, the sea deepened and formed a bed of limestone on top of the salt.

At roughly the time the limestone bed was forming, a mountain range rose to the northeast. As this eroded, vast quantities of sediment were carried westward from it by streams and dumped on top of the layers of limestone, gypsum and halite. According to a theory favored by some geologists, the weight of the new deposits produced an effect something like that of a large man slowly rocking his foot onto a toothpaste tube: over perhaps 150 million years, the salt and gypsum beds were squeezed upward through beds of the overlying strata. Here and there cracks and faults in those overlying beds allowed the salts to flow to the surface along paths of least resistance.

A thin layer of gypsum, deposited by ground water that seeped through nearby beds, slashes across a sandstone canyon near The Loop.

Today salt itself is almost nowhere to be seen at the surface. Much more soluble than gypsum, it has been carried off by rains or floods nearly as quickly as surface exposure or cracks in the rock permitted water to attack it. And with its disappearance, the salt cut the legs out from under parts of the canyon country. As ground water dissolved the underlying salt, the other rocks settled and buckled downward.

The rain had stopped by the time we reached Gypsum Canyon, and the April sun gave a thin warmth that vanished whenever a cloud hid it for a moment. The canyon floor was wide and easy to travel. Here the easily weathered gypsum outcropping had allowed the forces of erosion to create the broad valley bottom. Soft rock such as this typically makes for wide canyons, while hard rock results in narrow ones.

Just around the next bend in the canyon we came across the dividing line between the less durable strata on which we stood and the tough limestone of the Honaker Trail Formation, named for an old prospectors' route. The limestone had resisted the erosion of the canyon's stream, the softer stuff had not. The result was a difference in elevation of some 200 feet—the height of the beautiful cascade in front of us.

The falls came out of a notch far overhead that seemed completely inaccessible. The only way to continue was by climbing up the canyon wall till we were above the lip of the waterfall and hope then to find a ledge that would carry us upstream past the obstacle. But the canyon walls near the waterfall were unscalable; we would have to go back down-canyon till we could find a talus slope offering access to the higher reaches of the canyon.

Talus slopes are the ramps that make travel in the canyon country merely difficult instead of outright impossible. They are rubble heaps formed by the disintegration of cliff faces. Plant roots work into the cracks of the cliff, and water follows the same path. The water, freezing and thawing, leaching away the chemicals that hold the rock together, works the cracks wider until one day a pebble or a huge block tumbles free. The process can be slow, almost beyond human powers to perceive. Photographs taken a hundred years apart of the same cliffs and talus piles sometimes show little visible difference. At most a boulder or two will have moved a few inches downhill. But the centuries do pass. And the inches add up until each boulder eventually reaches the canyon floor, breaking up as it descends.

We found such a talus slope around a bend in the canyon, but the ledges to which it gave access looked, from below, to be hardly more than variations in color between strata of rock. These seemed unlikely to offer footing, and yet, if they did not, our exploration would end.

We started slowly up the talus slope—slowly because of its steepness, because the rocks were slick with rain, because the wind spat rain at us and turned our ponchos into sails that tried to yank us off the hill. Slowly, too, because we did not want to start a landslide, and slowly, at least in my case, because of fear. Once we reached the level of the most promising ledge, we could see only a section of it. The rest was out of sight around the bend, and we had no way of knowing whether it pinched out to nothing. The shelf where we already stood was close enough to nothing—just a narrow strip, wide enough for one man, with a slanting shoulder that dipped steeply downward for a few yards from the shelf and then dropped off sheer.

Pure cowardice got me onto that ledge: the fear of what the others would think if I did not dare follow them. I looked at nothing but my own feet, and moved them from one foothold to the next with the care of a tightrope walker. Fortunately Frank Nordstrom, in the lead, was paying better attention to the world around him. When he reached the turn in the ledge, he saw in front of him a band of nine bighorn sheep. These graceful, tough, agile animals, once common in the canyon and mesa country, are now rarities. "They eyed us warily," Frank wrote in his diary that night, "and finally trotted off as if the slope were a pasture —which it was, to them."

Not to us. But by going single file and hugging the cliff face, we were able to work upstream and to get back to the canyon floor once we had passed the falls. The process we used is called "contouring around" an obstacle. More informally, canyon hikers describe it as "going up the walls," an expression that gets the idea across admirably. By either name, the technique evolves from the fact that wherever a canyon wall is merely steep rather than vertical, the ledges formed by the basically horizontal rock strata will parallel the contour lines cartographers plot on their maps to link points of the same elevation. This particular ledge had allowed us to work our way upstream from the waterfall until we got to a talus slope gentle enough for us to regain the canyon floor. As darkness gathered, we made camp for the night.

Next morning, the going at first was easier. The stream had cut at a fairly even rate through the homogeneous limestone underfoot, so that the canyon floor sloped upward steadily instead of in a series of precipitous rises. Had we been walking the same route some 290 million years before, we would have been sloshing our way out of the shallow sea. From this marine limestone, part of the Honaker Trail Formation, we were climbing a few feet, and a few million years in time, out onto the mudbanks of the ancient shore, which had hardened over the eons into Halgaito shale. Where the two rock formations met, a series of waterfalls turned our hike into a struggle again. Once more we had to climb up to ledges to contour around each fall.

Going up the walls, the wise man follows not only contours but animal tracks. Most canyons in this region are narrow enough so that all large creatures in search of the best route eventually hit on the same one. And the other large animals, unlike us, had had all their lives to think about the problem. Along much of the way, we had been following mule-deer, coyote and bobcat tracks. We suspected some of the hoofmarks and droppings to be those of the bighorn band Frank had spotted the evening before. If we were right, they were working up the canyon not far in front of us.

Jim and Haskell were ahead, on the other side of a rocky rise, when they finally came across part of the band. They called out a warning to us as two of the sheep headed down-canyon toward us to escape. While Frank and I stood frozen, the two sheep—a mother and her partly grown offspring—came over the rise not 20 yards from us. Neither spotted us until we moved and then both fled up the wall in great pogo-stick bounds. If it is rare to see mountain sheep at all these days, it is incomparably rarer to see them at such close range.

As we went on, grappling, scrambling and slipping where the wild sheep had bounded so easily, our concern with geology was becoming a very intimate matter. To a man picking his way along a wet and sloping ledge with a hundred-foot drop below, knowledge of the rock under his hands and feet becomes crucial. And this rock was far from solid.

The canyon and mesa country, with few exceptions, is made of sedimentary rock—sand, mud and sea creatures, laid down as sandstone, shale or limestone by wind or water. Some limestones in watercourses become worn to such a dangerous slipperiness that they are aptly called slickrock. Sandstones may break underfoot or crumble when a man grabs them, and some are so soft that bees can actually tunnel into them. Shales are sometimes hardly more substantial than mud, baked by the sun in the bottom of a dried puddle.

At one point Frank lost his footing on a slick slope of red shale, glistening with water that made the rock's surface like grease. I heard his bamboo walking staff clatter on the rock, and looked to see him sliding for the drop-off where he would fall to his death. I was so sure there

The dark pigments of a substance called desert varnish streak the sandstone walls of White Canyon. This so-called varnish, found in all deserts, is created when water seeps through rocks and soil rich in manganese and iron oxides. The water dissolves these minerals, carrying them down the rock face, where they solidify into a thin film as the water evaporates.

was no hope for him that I remember wondering, in that second, how we would retrieve his body. But he saved himself with the instincts formed in years of hiking the canyons. He flattened his sliding body to the rocks so as to obtain as much friction as possible—and it was enough. He slowed to a halt a few feet from the lip.

At another season the shale that nearly did for Frank would have been easy to cross. But the winter still ending had been the wettest in memory. Even at the end of April, snow lay hip deep over the highlands and meltwater ran down to fill the canyons with torrents. Along the way the water had loosened the underpinnings of plants and rocks alike, so that the walls around us were eroding at what was, geologically speaking, a great rate. Several times we picked our way over fresh landslides —instant talus—in which juniper trees that had stood for centuries lay fallen and broken.

Over recent landslides like this, you move warily, hoping that the angle of repose has been reached. This angle, which varies according to the size and shape of the rocks comprising a slope, is the pitch at which the rocks will no longer shift or slide of their own weight. Until this angle is reached, the whole pile remains a potential landslide.

Not only are small piles of rocks such as these unstable in human terms, but rocks in the mass are unstable in geologic terms. Compared to a big toe, for example, they may seem very solid and substantial indeed. But large rock masses can, in reality, be fluid—as in the case of the salt structure at Gypsum Canyon. In addition, many rocks are porous, allowing liquids to pass through them with comparative ease.

Oil, for example, seeps through rock and collects in what are called pools, although they are not pools in the ordinary sense. They are places where the petroleum has soaked into not-so-solid rock—very much as motor oil will permeate the concrete of a filling station apron—and then has become trapped by a layer or layers of less permeable rock. If water is trapped in the pores of the same rock, the oil will gradually rise until it floats on top of that water. Thus, not only are the rocks themselves fluid over the long, slow haul, they also act as conduits and containers for other fluids.

As we ascended, we left the Honaker Trail limestone for the Halgaito shale. In so doing we gradually traded steep walls and waterfalls for the wide bottomland of upper Fable Valley. Here were grass and groves of trees, and room to look on all sides. The land was accessible; its owner, the Federal Bureau of Land Management, rents it to cattle

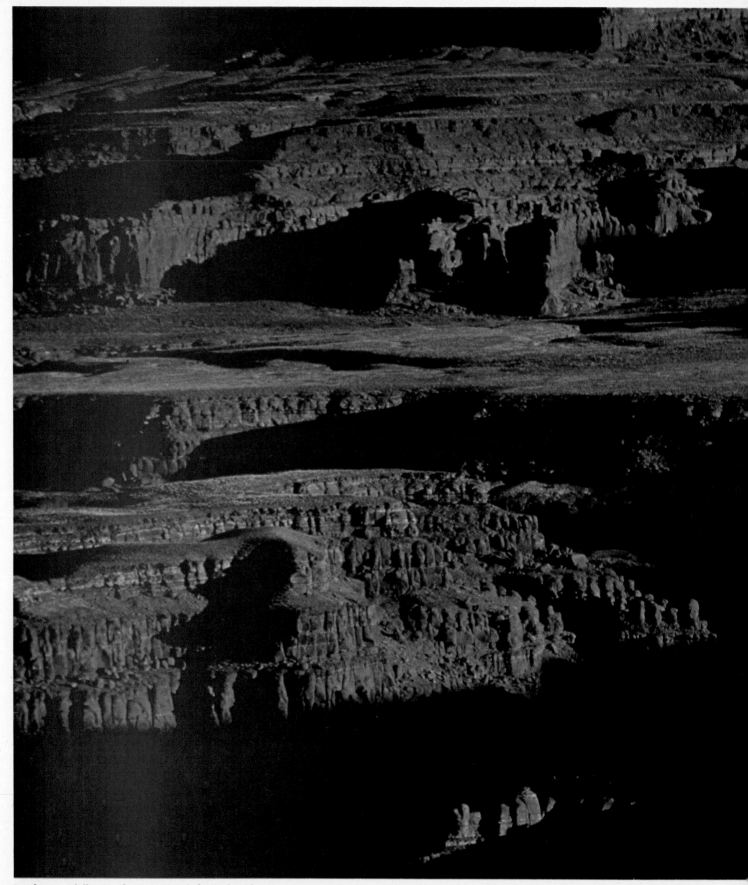

Sandstone cliffs rim the canyon of the Colorado River (center) in this sunset view from Dead Horse Point. The name recalls the fate of a

herd of wild ponies that were rounded up and culled for breeding stock by cattlemen, then turned loose—subsequently to die of thirst.

ranchers—and so here, too, were cows. A handful were grazing a quarter of a mile away, in a grove of Gambel oak. Grazing with them were three mule deer, which took alarm when we came into view. They sprang toward the safety of the mesa and in an instant were on the valley wall. Their path was zigzag, and at each switchback they paused for a second like skiers surveying the next stretch of trail. Then they went on, leaping across the hill and upward, never missing their footing. Very quickly they disappeared.

A little farther into the valley we came across the first sign of man himself—the tab from a beer can. We soon crossed the actual track of the beast: two parallel lines pressed into the soil. We followed these Jeep tracks along the flat bottom land before leaving them to head for a spring on the slope of an oak-covered bluff. From there we could see, halfway across the valley, a steep cockscomb of rock on which sat an ancient Indian watchtower. A small stream had already cut a trench some 15 feet deep through the alluvial material that filled much of the valley bed. In time—perhaps well within our lifetime—Fable Valley may thus be scoured entirely clean of bottom land and return once more to a rocky, typically V-shaped canyon. Such things have happened in the recent past, and are happening today, throughout the canyon country. Toward the middle of the 19th Century, Mormon pioneers found fertile meadows and even little ponds in canyons that today are stripped nearly bare of the soil that once formed their level floors.

Exactly what causes canyon bottoms periodically to fill with stream-borne sandy deposits (aggradation) and then be scoured empty again by floods years later (degradation) is a matter of conjecture. During degradation, the stream cuts a valley flanked by long terraces and miniature palisades into the sandy soil along its course. Year by year the waters cut these terraces farther back from the stream bed until most of the canyon is swept free of alluvium. As more years pass, slight changes in rainfall may tip the balance either way, and the streams mysteriously begin to fill their canyon bottoms back up with alluvium.

Whatever happened to trigger the current cycle took place in the very recent geologic past. One Navajo legend blames the disappearance of fertile bottom land on a curse pronounced in 1884. Curse or not, meteorological records from the period show that there were indeed abnormally heavy snowfalls in 1884 and 1885, leading to unusually damaging spring floods in those years.

In the rock of Redbud Pass near the Colorado River I have seen a souvenir of such a land-building cycle. Fifteen feet up the canyon wall was a frieze of fossilized roots, in perfect detail down to the smallest root-lets, running for yards in both directions. Thousands of years earlier, during a time when the canyon had been filled up to that level with sed-iment, the roots of trees and bushes had sought out water seeping from the canyon wall, and had spread underground along its face. When the roots died they were imbued with calcium-bearing water and the whole network was exposed, fossilized on the face of the wall.

That evening in Fable Valley our main concern was warmth. Away from the circle of our fire the temperature dipped below freezing while we ate supper. By the next morning, hoarfrost covered our sleeping bags and the water in an 11-quart plastic jug had frozen nearly solid. We loaded up and moved out quickly, on the theory that exertion would create warmth even if the sun did not. We began to run into deep snow on the shaded, north-facing slopes at the head of Fable Valley, where the Halgaito shale gave way to cliff-forming Cedar Mesa sandstone. To leave the valley we had to climb through piñon pines up some 600 feet of giant stairsteps formed where sandstone alternated with softer shale strata. In warmer weather these ledges might have been so high and steep as to have made our route impossible; now we could climb be-cause snow had drifted deep between the steps and crusted to make a huge ramp. A lone coyote had been up before us; his tracks showed us the best path to take.

Once on top we were near the 8,000-foot mark, and snow lay waist deep over the Elk Ridge Plateau. Most of the mesa top was open coun-try, although piñon pine and juniper grew here and there. The snow cov-ered the sagebrush, which trapped air pockets within its branches so as to form pitfalls. When we crashed through the crusted snow, we had to flounder free the best way we could, fighting the brush as well as the snow. But mostly the crust was strong enough to support us.

By midday we had reached the edge of the plateau, a 1,000-foot, sheer cliff of sandstone. To look out and down was to look from one sea-son into the next: the springtime world below us stretched snow-free, green, gray, red and tan to the horizon. To our north, Dark Canyon, the principal drainage for the Elk Ridge Plateau, ran off westward toward the Colorado River. Into Dark Canyon plunged Youngs Canyon, Lean-to Canyon, Black Steer Canyon, Lost Canyon—all formidable clefts in their own right. The excavation work represented by these canyons is impressive enough. But of far more significance to the geologist is what cannot be seen—what is simply not there any more.

We were sitting on the rim of the sandstone cliff, at an altitude of about 8,000 feet. Some 1,500 feet below us, cut by the complex of canyons, lay the more or less level landscape at the base of that cliff. In between was nothing: where now we looked through clear desert air to the horizon, all had once been solid rock. Throughout the Colorado Plateau, the real monuments to erosion are not the remaining spires and buttes (of which those in Monument Valley are perhaps the most striking examples), but the vast empty spaces that separate them.

At present the empty space barred us from further progress toward Dark Canyon, a broad fissure in the plain far below. But the map showed a westward-trending tributary of Youngs Canyon that seemed to offer us a way down. In less than an hour we had reached the beginning of the nameless tributary canyon. At first the snow was drifted deep and we kept plunging through the crust up to our hips. After a time, however, the south-facing slopes became nearly dry, although the snow remained deep on the north-facing side. But even this vestige of winter diminished with every step as we descended farther into the springtime.

After losing about 300 feet of altitude, we came around a corner to find our way blocked by a 200-foot drop-off where the sandstone plunged precipitously to meet the underlying and softer shale. But around a shoulder of the canyon we found a joint—a natural fracture in the rocks down which tumbled a talus ramp leading across the cliff. Such cracks can be formed in a number of ways—by the arching of the rocks beneath, by the subsidence of those same rocks, by any shifting of the earth's crust. In these joints, or fractures, erosion begins its work. This particular joint, given enough millions of years, might cut its way back into the highlands behind to become a tributary canyon itself.

There is little or no indication on even the best maps where such access routes may be. Maps fail to distinguish the passable talus slope or cliff from the impassable; a fissure in the rocks does not show at all. At best, contour lines indicate altitude differentials of either 40 or 80 feet. If all or most of that is sheer drop, the hiker is blocked. The little blue lines on the maps—indicating the possible presence of water—are deceptive, too. Sometimes they will represent a series of small, clear pools. At other seasons, in the same canyon, they will mark only the place where water once was, and for the hiker nothing but dry sand whispers through his fingers if he digs for water in the stream bed. At yet another season the blue lines will stand for surging brown torrents, often so heavy with rocks, sand and mud as to amount to liquid landslides

On a canyon rim near the San Juan River, a desert bighorn stands above a growth of mountain mahogany, a favored browse. Though bighorn rams may weigh as much as 200 pounds, they can bound with ease from slippery shelf to narrow ledge because their sharp-edged hoofs have a rubbery interior surface for gripping the rock.

hurtling murderously down the canyons. Our next day was to make a number of these points.

We had passed the night on the nameless tributary. Its junction with Youngs Canyon, to judge by the map, could be either impossible or relatively easy. It turned out to be something in between: a descent that looked impossible from the top, but proved to be feasible once we had worked our way down the rock face a little. The blue line running down the tributary canyon marked the course of a small stream that was actually invisible for much of its length; it seeped along underground, surfacing here and there in a series of rocky pools. Youngs Canyon was much the same at the moment, but not many days before the scene had been very different. All around were signs of an awesome flood: driftwood jammed in the rocks well over our heads; large pebbles stuck in the bark fissures way up the trunks of the cottonwoods; the corpse of a mule deer that had been caught in the torrent.

After little more than a mile, Youngs Canyon ran into Dark Canyon, whose stream was neither clear nor intermittent. It was wide and turbulent, brown with mud. All along the way, the sandstone and limestone of the Honaker Trail Formation were full of fossils—sometimes dozens to the square inch. Algae and the shells of tiny marine animals were preserved in a brick-red rock called chert. This is a siliceous material much like flint, enough harder than the surrounding limestone so that the little fossils were raised in bas-relief on the gray, eroded surface of the rocks. Often we came across large nodules of chert embedded in solid limestone like flinty cysts. These are formed, the geologists think, from silica carried to the oceans in streams. When the fresh stream waters entered the salt sea, the silica came out of solution, often in soft globular masses that eventually were engulfed by the formations of limestone on the ocean bed.

In Dark Canyon we also came across another dead mule deer. The floods just past must have moved with enormous speed to have caught animals so strong and agile. And now it was raining again. With the deer in mind, we kept an eye out for ledges high enough to give us refuge if the floods returned. Once, the noise of thunder came down the canyons toward us, rolling and rumbling back and forth between confined walls, growing louder, breaking over us with a crash, and then dying away in a long series of ever weaker echoes. But a hard storm never came, and the water level held steady.

That night we camped near the junction of Dark Canyon and Lean-to Canyon. Our shelter was a niche not a yard high running along the bottom of the canyon wall. Each time I brushed the rock roof above me, grains of sand pattered softly down. Water had slowly leached away the mineral cement that had bound the surface grains together into solid rock. Where the ground water had evaporated on reaching the rock face, it left behind a bloom of white gypsum.

Outside the rain was coming down harder, but not hard enough to put out the fire we had made of juniper wood. We lay dry in our low alcoves, smelling the perfumed juniper smoke and listening to the raindrops plop and hiss into the fire. Next morning the rain had stopped, although fog still hung in the air around the tops of the red pinnacles and spires surrounding us. Something about the sight tugged at the hem of memory. Something seen in childhood that I could not quite bring back. The steel-mill smoke of Pittsburgh clinging to the red brick turrets of some vast Victorian pile? A post office? A railroad station? Gone now, whatever the memory was.

When we reached the top of the precipitous 800-foot talus slope leading out of Lean-to Canyon, we found that six inches of wet snow had covered the mesa top during the night while rain had been falling in the lower altitudes of the canyon itself. We tracked through the already melting snow toward the place we had left our truck five days before on a cattleman's Jeep trail. The trail took us to a paved road that led eventually, like all roads, right back to Pittsburgh. But as long as there remains at one end of the road a Lean-to Canyon we are still in reasonably good shape. When there is a Pittsburgh at both ends of the road, then there will be no place to go at all.

A Natural Architecture

The stony terrain of the Canyon Lands reveals some of the most astonishing varieties of shapes on earth. Here are forms and structures that seem less the random work of nature than the massive artifacts of human architecture—towers and domes, walls and buttresses, bridges and pillars, and even a few formations that suggest human figures.

Yet paradoxically, these forms in the canyon country are not so much a result of building and creativity as of tearing down—the destructive work of rain, snow, river water, frost and wind. They have been left behind as the primordial landscape has disintegrated under the inexorable cycles of weathering and erosion. Now they stand like the towering ruins of enormous and abandoned cities, striking the eye of the wilderness visitor with a force that seems to demand description. Such, indeed, appears to have been the experience of the original Spanish explorers who first beheld the landscape. Ultimately they found words to describe their specific impressions of the various rock forms, words that were as suggestive and full of imagery as the canyon country itself.

"Canyon" is of Spanish origin —possibly derived from an old word for street, recalling the twisting, steep-walled thoroughfare in a town of tall medieval buildings. *Mesa* is the Spanish word for table and once again the imagery describing the massive, flat-topped platforms of rock that rise abruptly from the desert floor is both appropriate and clear. No less graphic are English words applied to other forms: spire, bridge, staircase, pinnacle, needle, monument, and a maverick albeit bluntly intriguing word—badlands. The French word *talus,* attached to the slopes of broken rock that are found at the bottoms of cliffs in the canyon country, once referred to a more romantic image—a hillside harboring gold under the soil.

Though the labels given to the canyon-country land forms vividly communicate an idea of their shapes, few of the names even hint at the extraordinary spectrum of colors —muted by snowfall in some of the pictures on the following pages —that are splashed everywhere in the region. Only the Spanish *colorado,* conferred by the area's first Old World explorers on its principal river, conveys an idea of the land's brilliant palette; the word, meaning red or reddish, truly matches the region's most dominant hue.

Arranged like the ruined walls and towers of a medieval fortress, sandstone buttresses dominate the background of a scene on the Colorado Plateau. These walls—once the sheer ramparts of a larger mesa—have gradually broken up along vertical fractures in the original formation.

Isolated on the Kaibito Plateau near the Arizona-Utah border, a typically flat-topped, steep-sided mesa rises from the desert floor. Massive though it is, the mesa is only a relic of the unbroken covering of sedimentary rock that once stretched to the horizon.

Deeply carved by a small tributary of the Colorado, White Canyon wends its way through a narrow gorge of nearly vertical walls several hundred feet deep. In this winter scene, the riverbed, virtually dry most of the year, is masked by thin drifts of snow.

Castle Rock, a sandstone pinnacle in Utah, rises like a watchtower from its foundation of softer layers. Along with the satellite block at far left, Castle Rock represents a late stage in the erosive destruction of a mesa.

A stretch of badlands, devoid of plant life, sprawls beneath early snow near Caineville, Utah. These low hills, made of soft, heavily eroded

shales and siltstones, are an anomaly in canyon country, where the more prevalent formations of hard sandstone and limestone resist erosion.

Pierced by a canyon stream, Hickman Natural Bridge spans a dry gulch in Capitol Reef National Park, Utah. This form was once a solid rock jutting out near the junction of two canyons. At first, the river swept around the obstacle, but slowly the water undercut the rock and tunneled through. Weathering and erosion enlarged the new channel into this graceful arch.

Layers of sandstone in Arches National Park acquired their resemblance to a ruined staircase from a process called stripping. Throughout the year, ground water and rain seep into the porous rock. Winter temperatures freeze this moisture, which expands and cracks the rock into fragments. Desert winds then blow away the loose particles, baring the chipped strata.

Monument Valley's Agathla Peak, also
known as El Capitán, is an enduring
mass of unusually tough rock called a
volcanic neck. Once a roiling pool
of molten magma that filled a volcano's
throat, the fluid thickened as the
volcano became dormant. Cooling, it
solidified into a hard mass highly
resistant to erosion. Eventually, its
encasing outer layers wasted away to
leave the interior neck standing alone.

Basalt boulders, another volcanic
remnant, pepper the high ground in
Capitol Reef National Park. Wrenched
by weathering from mountains some 15
miles away, they were carried by
long-vanished rivers to the foot of the
domed peak in the background.

Pebbles and boulders of a talus slope —the mingled debris of a disintegrating cliff—form a bank of the Colorado near Moab, Utah. Unceasingly, this ragged slope creeps down to the riverbed, moving barely an inch or two a year, yet steadily making way for new rockfalls from the cliffside.

3/ The Forbidding Country

The closer one gets to the master stream and its major
tributaries, the deeper the downward departures
become. The canyon predominates. Not one canyon,
but a thousand.   C. GREGORY CRAMPTON/ STANDING UP COUNTRY

From the Valley of the Gods a dirt road scrabbles its way some 1,000 feet up Cedar Mesa in a series of switchbacks. When the road levels out on top of this immense mesa in southeastern Utah, a smaller dirt road forks off and runs five more miles down to Muley Point, on the northwest rim of the mesa. From there it is only a few yards' walk to a panoramic view of one of the grandest yet most forbidding stretches of country in the world.

Rising and falling in dramatic sweeps and clefts, the soft blues, reds, yellows, oranges and greens of terraced rock go on until they fade into the violet haze of the far horizon. In the near distance is the austere slash of San Juan Canyon, whose river is so powerful in floodtimes that it has chewed down as much as 2,000 feet below the rimrock but nevertheless several times a century dries up altogether during the extended droughts. During the great drought of 1896, it could be crossed dry-shod; and the same was possible again in the exceptionally dry summer of 1934. Beyond the San Juan to the northwest, blocked from the line of sight by distance and by masses of intervening rock, are three of the San Juan's major tributary canyons: Slickhorn, Grand Gulch and Johns Canyon. These are the canyons that I know best. They, too, suffer from the same cyclical want of water as the San Juan and the whole intervening country. All three are watered principally by runoff from the Bear's Ears, two 8,000-foot-high rock formations lying about 25 miles away. The three canyons thus receive a pretty fair supply of water for this corner of Utah, but their streams nevertheless dwindle or disappear during the dry parts of the year.

From the overlook at Muley Point, hardly a trace of vegetation appears anywhere on the land below. There are no buildings and no paved roads traverse the country. In the hundred or so square miles lying between Johns Canyon and Grand Gulch, only three Jeep tracks wander over the uplands. One of the tracks intersects Johns Canyon and years ago, before the roadway fell apart, reached as far as Slickhorn. But today, only the upper part of Johns Canyon can be reached by motor vehicle. Slickhorn, Grand Gulch and the lower portion of Johns Canyon can be traveled only by foot or on horseback.

The spectator at Muley Point gets the impression that no human could —or would—inhabit a wilderness so bleak and barren. And, insofar as no human being does live there at the moment, the spectator is correct. Yet, in the past, people have tried. The U.S. Geological Survey maps of the region show 19 Indian ruins. And I know, because I have been there and seen it, that at least one ruin—the foundations of a stone shelter in a nameless side canyon—does not appear on the maps. Wood, metal and plastic remnants in three other locations show that white men, too, once lived in the rough rectangle bounded by Johns Canyon, the San Juan River and the giant dog-leg described by Grand Gulch.

The thing that first brought the white man was oil. In 1879 a man named E. L. Goodridge found "little streams of oil coming from loose boulders" just up the banks of the San Juan from Slickhorn Canyon. Goodridge staked a claim near the present site of Mexican Hat, Utah, but 28 years passed before enough money was collected to go ahead with drilling. In 1908 the wildcatters brought in a 70-foot gusher, and other oilmen rushed in to follow up on the discovery. Wells were drilled in the San Juan country off and on for the next 20 years. But the oil seemed to occur only in small pockets, so that hitting it was primarily a matter of chance, and most searchers came away after much effort totally empty-handed. Only a half dozen or so of the holes ever amounted to much. And even their production was so small that most of it was used up in a curious sort of economic closed circle: in a country where firewood was nearly nonexistent and coal too expensive to haul in, the oil from the few producing wells was used to fuel the rigs that drilled the many dry holes.

In 1938, after the boomlet was over, government geologist H. D. Miser made a tour of inspection into the area and reported: "The oil comes up as bubbles in the water (of the San Juan River) and as minute streams through sand and also through boulders of sandstone and limestone at the edge of the water. . . . The drops of oil break on reaching the surface of the river and spread as thin iridescent films. . . . At one seep a film of oil covering several square feet of a sand bar is used by flies as a breeding place, and hundreds of larvae live in the oily substance on the bare surface of the sand bar."

And so the San Juan country turns out to contain oil in exploitable quantities after all, but only if you are a fly. The canyon rivers themselves, it turns out, are in part responsible for this. By gradually cutting down through the rock layers where oil is normally stored, they have over the eons given it a chance to seep away, as it is slowly doing today in the canyon of the San Juan.

The region had its gold rush too. In 1891, rumors began to circulate that the terraces and sandbars along the San Juan River were rich in gold. Nearly every miner's pan did show color; but the tiny flecks had probably been washed downstream from the San Juan Mountains of neighboring Colorado where the rock strata do contain small amounts of gold. Although most of the 1,200 prospectors very quickly learned that the gold existed only in quantities far too minute to be profitable, some hope persisted, and there were bursts of mining activity along the river well after the initial boom had bust. However, by 1912 even the most optimistic had given up. The only enduring souvenir of the gold rush today is the Honaker Trail.

In the whole San Juan country the single prospecting venture that proved to be worth the effort arose more than half a century after the gold rush, with the discovery of the atomic bomb. During the early 1950s, deposits of such uranium-bearing minerals as carnotite suddenly became convertible into yachts, fur coats and any number of pleasant things. In the age of the Geiger counter, prospectors were limited neither to such primitive tools as pickaxes nor to such plodding transport as mules. One favorite technique was to fly along cliff faces in a small plane with a sack of flour ready to toss out the door wherever the counter aboard gave a promising reading. The sack would burst on impact, leaving a white splotch to guide the prospector when he later returned on foot to collect samples.

Again white men crowded into the Canyon Lands. A few prospectors made fortunes, but then, partly as a result of their endeavors, the nation's uranium stockpiles grew larger and the price of the mineral

Utah junipers stud the plateau above an arroyo that winds between buttes before plunging into the abyss of Taylor Canyon (far left).

A mule-deer buck, lured to this canyon bottom by patches of grass, senses the photographer's presence and prepares to flee.

dropped. The canyons returned, not too badly scarred, to the tranquillity that had been theirs for centuries.

Grand Gulch, Slickhorn and Johns Canyon were barely touched by any of the prospectors. Yet a millennium earlier, Indians inhabited all three, at a time when Grand Gulch, in particular, must have had a relatively dependable supply of water and plenty of cultivable bottom land. All along the upper reaches of the canyon are ruins left by a mysterious tribe called the Anasazi, who moved into the Canyon Lands a thousand years ago, and then disappeared sometime during the 13th Century. Where they came from and where they went is unclear. However, they left behind them hundreds of dwellings set into caves or under overhanging ledges on the canyon walls, high enough to be out of the way of floodwaters as well as enemies, and usually facing south for warmth in the winter.

In a land so empty, to come across traces of the vanished Anasazi seems endlessly strange to me. As a practical matter I know the Indians were once there, just as the second astronauts on the moon knew that the first had preceded them. But I have always thought that the actual footprints in the lunar dust, had they landed in the same vicinity, would have come as a shock. I feel the same surprise about the Anasazi, whose footprints are gone but whose handprints and other vestiges are common. A vast natural arch spans Grand Gulch at one point, making a sheltered campsite used by the Indians so often that saucer-shaped depressions were worn into several flat-topped boulders by women grinding corn into meal. A few yards away, hundreds of red and black handprints decorate the rock wall. Did the Anasazi smear their hands with paint and leave these marks on the stone as some sort of tribal census? No one is left to say.

That no one is left seems curious, too, upon first glance; many Anasazi ruins give the impression of having been vacated only a few years ago. The modest shelters of the Anasazi have lasted through the centuries in an extraordinary state of preservation. The construction was crude but durable—blocks of sandstone cemented together with a mortar made of dried mud. The sites were often in niches, sheltered from the weathering effects of rain and snow. The wooden roof beams and doorframes are preserved, too. Wood attains something close to immortality in the desert, where the air is so dry that rot-causing fungi cannot thrive (radiocarbon dating done in 1970 by C. W. Ferguson of the University of Arizona on a piece of driftwood from a cave along the Colorado River showed that the fragment was more than 35,000 years old).

One day I climbed up to such a ruin in Grand Gulch and spent half an hour poking around, musing. Shards of pottery lay all around in the dusty sand. But I was not surprised that these had proved so durable, since I knew that pottery from ancient Mediterranean civilizations has lasted much longer. What did astonish me, though, was to find the site littered with corncobs, as well preserved as if they had been gnawed clean only last week. I went inside one of the sandstone buildings, a low, small structure that was partly below ground and might have been a root cellar or similar storage area. There I found pieces of gourds, squash stems, part of a basket woven from some vegetable fiber, and a curved wooden pot handle. The dust that lay thick on them was the only sign that they had been lying inside that Anasazi structure since the time of Genghis Khan.

Beside the storage building were the remains of some sort of circular edifice, its floor also slightly below ground level. Many archeologists believe that such places were used for religious ceremonies, like the similarly shaped *kivas* of the Pueblo Indians. I, on the other hand, believe that archeologists have a tendency to explain away anything they don't understand by saying it was used for religious ceremonies. I prefer to think that these circular depressions were actually the remains of steam baths, communal and probably coed. In holding this steam-bath theory, I find I am not alone; two archeologists, Gordon Vivian and Paul Reiter, agreed with me in a 1960 study—though they insisted that the baths were for ritual cleansing.

I am possibly the original discoverer of another such circular ruin, although my find remains unaccountably uninscribed in the annals of archeology. One spring I was exploring a side canyon off Grand Gulch when I came across a large and elaborate Anasazi mural. The painting stretched along the wall for some 10 yards—snake designs, sticklike figures representing deer, arrows and hunters, and various abstract patterns. To one side were foundation stones outlining a vanished structure that had been oval in shape—another steam bath, I had no doubt. I knew the chances were that I was not really the first nor even the hundredth to come across the place—the side canyon was not particularly difficult of access. But though the U.S. Geological Survey tries to mark both Indian ruins and paintings on its maps, it had not marked these; so who was to say I was not the first? Besides, I *felt* like the first to see the ruins and the mural.

The beetling 1,500-foot-high walls of Clearwater Canyon enfold stream-bed hackberry trees in almost permanent shade.

This aerial view reveals the intricate shapes of the rocks in Canyonlands Park's so-called Maze, whose contours are striking examples of the formations throughout the area. The indentations along the canyon walls indicate the presence in the sandstone of fractures caused long ago by the earth's shifting crust. The original sharp angles of the separated sections have been rounded by weathering and erosion, and some parts of the eroded sections have split from the walls completely, tumbling to the canyon floors.

Anasazi art takes two forms: pictographs and petroglyphs (the first are painted on the rock face, the second chipped into its surface). Some scholars feel that the designs are accounts of hunting trips left for the wonder, envy or instruction of future hunting parties. This sounds logical, but despite the fact that the colors used by the long-dead Anasazi artist in my particular discovery were still bright on the canyon wall, the complete meaning of his message was lost on me. I did get a plain impression that the mural was not merely decoration. It told a story, yet the painting was like a tale written only in nouns: the hunters, the arrows and the deer in the picture were obvious enough. However, the abstract designs and markings meant nothing to me.

Scholars cannot read the story clearly either—they know very little about the Anasazi. The name is a Navajo word, usually translated as "the old ones" but sometimes interpreted as "the ancient enemies," a meaning that may reflect mutual hostility between the ancestors of the Navajo and the Anasazi. From roughly 900 to 1250 A.D. (dates proved by tree-ring and carbon-14 dating techniques), the Anasazi were in the Canyon Lands—and then they were there no longer, driven from their canyon homes by war or a prolonged drought. Disease may have swept the tribe, although the Anasazi settlements were widely scattered. Perhaps a combination of these things weakened the Anasazi and caused them to disperse. But where did the survivors go? Again, no one knows —or may ever know. However, archeologists speculate that the Anasazi might have joined the ancestors of the Hopi, who now live more than 100 miles to the south in Arizona, and may have become culturally submerged in that civilization.

Near the ruins the little side canyon was watered by a clear creek running through reeds as high as a man. Once this water might have irrigated Anasazi crops. Today the stream and the shelter from the elements offered by the canyon walls have created and preserved one of the beautiful natural anomalies of the canyon country. Where conditions are right, as here, occasional pockets very like a New England deciduous forest occur, tucked down into the heart of the desert. The various plant species may be slightly different from those found in the Eastern United States, but the genera are the same: oak, Virginia creeper, poison ivy, Juneberry, columbine, dogwood, false Solomon's-seal and violets. The trees are diminutive Gambel oaks, with leaves only a couple of inches long, yet unmistakably oak leaves. They have the familiar rounded lobes and the look of dark-green polished leather.

Unlike larger oaks, however, Gambel oak grows in groves whose individual trees often come up from a single root system and they are therefore extremely durable. Burn or otherwise destroy one tree, or two or a dozen, and you have not destroyed the system. New shoots simply come up again, so that Gambel oak is one of the first trees to reestablish itself after a fire.

This little grove set down into the desert had the smell of the canyons, but that is to explain nothing, really. I can only say that canyons, in general, smell good. To find out in just what way, you must go there and smell for yourself. In this side canyon off Grand Gulch, for example, the smell was of dryness. Dryness has its own smell, here compounded with that of the dead oak leaves underfoot and the thin soil to which they would someday lend their substance as they decomposed. There is a half-dry and a half-wet smell, too, when fat raindrops first start exploding in the dust with little puffs. And a wet smell, when the thunderstorm has passed. There are flower smells as well, that are different with each season. Often you will be walking in the canyons and suddenly become aware that the air around you has been perfumed for some time. Perhaps you will see the flowers that make the perfume; perhaps they will be too small to notice, or too far away for the source to be apparent.

There is also the smell of sage; this you make yourself by forcing through the bushes, bruising leaves so as to let loose their scent. There are the smells of dry, dead juniper and live juniper, and of juniper smoke, of the amber pitch bleeding from piñon pines, and of the mariposa lily and of a hundred other things.

Nearly all of the hundred smell good, too; there are few bad smells in the canyons. Rottenness, mold, decay—all these require constant dampness to reach their full measure of stinking corruption. In other places, wet is negative—clammy or cold, suggesting decay or spoilage or angry hens. But in the Canyon Lands, wet is positive. I have sometimes stopped and scratched a hole into a patch of damp sand below a dry waterfall to see if water would seep in. Not that I needed the water at the moment; I only wanted to see if this was a place I could mark on my map and come back to, just in case.

Even when water is plentiful, as in springtime, it seems far more precious and beautiful than it might elsewhere. The springtime stream in Slickhorn is small and perfectly clear. It makes a succession of quiet pools, sun-dappled through the new cottonwood leaves. In the upper part of the canyon the stream runs, sometimes lost to sight, through and under giant boulders. Farther down, it slides gently from one pool to another over blue-gray limestone. Long, shallow scoops show in the limestone depressions where the water lies in long bathing basins. Along the way are tiny waterfalls, emerald green from the algae that cling to the lip of rock over which they tumble. I have even seen a fountain bubbling out of a hole in the limestone. The water there came from underground, with enough pressure to force a tiny, sparkling column four or five inches into the air.

Slickhorn Canyon ends in a little flood plain whose sandy soil is covered by a stand of grass. Near the creek bed are tangles of silver driftwood. Off to one side of the grassy flat, brush partly hides another human vestige—the ruins of a giant machine, colored softly by rust so that it nearly matches the red of the surrounding rocks. This old steam engine, once used to power a drilling rig, marks the farthest advance of technological man into the area. It is the tip of an inquiring tentacle that fell off and stayed, forgotten, to crumble gently back into its constituent elements.

A steep but short scramble up from the old engine, there is a flat place. Two rust-brown pipes stick up from the flat, only a few yards apart. One sits on top of a 267-foot well drilled by oilmen early in this century. The other pipe, marked "Dan Danvers No. 1," caps a 700-foot hole drilled and abandoned in 1952. According to the old drilling report, the wildcatters blew 300 quarts of nitroglycerine below ground to "straighten" the hole—once a common practice to break up underground rock formations and stimulate oil flow. But the technique, for whatever reason, did not work; unspecified "mechanical difficulties" developed and the well failed.

The debris still lying around on the rocky shelf—tangles of cable, a plank bunkhouse lying on its side—must all be from the earlier, unnamed well. By 1952, the time of Dan Danvers No. 1, drilling rigs could be hauled in and out of remote locations by tracked vehicles. However, when the first well was drilled, probably around 1908, the cost of hauling out the heavy equipment would have been considerably higher than what the stuff was worth.

Certainly the work of hauling it in must have been prodigious. First a road had to be built from Johns Canyon down the river to Slickhorn —all, in those prebulldozer days, by dynamite and pick and shovel. The terrace, far above the river along which the road ran, was flat enough, but the road builders had been forced to detour around the

Displaying distinctive yellow spots, greenish body and the black neck bands that inspire its name, a collared lizard basks on a sandstone ledge. Pugnacious, greedy and extremely active, the collared lizard is a predator that feeds on young snakes, small birds and sometimes on other lizards.

heads of eight small, dry canyons along the way. This meant that a straight-line distance of about 11 miles became 17 miles of road, nearly every yard of which had to be cleared of boulders and rocks of all sizes fallen down from the mesa above. The worst part must have been the last steep, rocky pitch from the high terrace to the lower shelf, where the well was to be drilled.

In the 20 or so years since it was last used, this sharp descent has already become completely impassable to motor vehicles. The places where men had built up the roadway have eroded away and fallen rocks again bar long stretches. One of these rocks, quite near the two wellheads, has a flat side about six by four feet, facing the trail and freckled all over, as neatly and clearly as in some educational display, with hundreds of small shells and other marine life. Remnants from the long-vanished inland sea, the fossils are set into the flat, gray surface in brick-red chert.

On top of the steep pitch, where the abandoned road begins to run flat along the cliff terrace, lies a scattering of boards and debris, including a shopping list and a single plastic clog-heeled shoe of the sort with which women deformed their posture in the early 1950s—and with which they have lately taken to deforming their posture again. Nearby I found an old spiral notebook full of practice business letters in shorthand. "Dear Mr. Barker: In reply to your letter of May 25th, we have the pleasure of informing you. . . ." A tough woman, to have lived out here in this shadeless, treeless, rocky place on the stepping-stone between two sheer cliffs.

They had to have been tough, those few people who made homes or searched for mineral treasure in this corner of the Canyon Lands. The prospectors, the oilmen and the Anasazi Indians all were tough, and clever, and brave, and persevering. But the land was tougher. Indifferently, the land tossed the men aside and held onto its treasure —which was never in the rocks anyway. The real treasure is the rocks: their massive shapes and subtle colors, their gorges and amphitheaters and secret fissures. The Canyon Lands will let you look at these things as a visitor, but they will not let you stay.

One morning in spring I was going down Slickhorn Canyon just as the cottonwoods were coming into leaf. I was hiking with some other people, but I had gone out ahead of them while they worked their way along a deer trail up on the canyon wall. I was down in the stream bed, which was piled full of boulders big enough that they were solidly settled and firm underfoot. The sun was just hitting into the canyon. As I went by a rock, a little white butterfly flew out from under the edge, took awkwardly to the air and then fluttered away. The time must have been right for the butterflies to wake up; in a few minutes others were dancing all through the air. Some of last year's leaves still hung on the cottonwoods, dabs of silver, dun or gold as the sun hit each differently. The new leaves were the palest green, not yet big enough to dance in the breeze like the old ones.

I began to run down the steep stream bed, just to use the way I felt, leaping along the tops of the uneven boulders. I made it into an athletic event, a kind of mountain sheep's steeplechase. The pack on my back was part of the game, there to make it hard to balance right while bounding along. The trick was to work the pack into the body's rhythm, until both were swinging smoothly together down the canyon like a falling leaf, never a stumble. Because, just this once, every movement slid easily and gracefully into the next the way they sometimes will on a perfect downhill ski run.

I pulled up out of breath at last by a still little pool. Last autumn's leaves on the bottom made the water look the color of pale tea. I took off my boots and socks and put my feet in the pool for just a moment; the water was so cold my feet quickly began to hurt and went numb. After a time the other hikers caught up to me and we went on together, now climbing down carefully over the huge, irregular boulders and taking our time like prudent people.

I hadn't found any gold or uranium or oil among the jumbled rocks of Slickhorn. But then I hadn't been looking for any.

NATURE WALK / Down Johns Canyon

PHOTOGRAPHS BY DAVID CAVAGNARO

I know a man whose favorite canyon is Johns Canyon, down in southeastern Utah. I know another who thinks the first man is out of his mind. "Nothing really special about Johns," the second man says. "Slickhorn is a better canyon."

I don't really have a favorite. My preferred canyon would be a patchwork assembled from memories: a looming cliff from this place, a fragrant patch of sage from another, a waterfall from a third, a narrow stream making its way past weathered boulders in a fourth.

Still, the man who preferred Johns Canyon kept telling me about the place until I began to think I had better have a look after all. Actually I had seen part of the canyon already —the upper part, near where an old Jeep trail fords the canyon stream. Above this point I knew—because I had once followed the trail—that Johns Canyon was wide and grassy, grazed by cattle and horses, and that it was the site of an abandoned homestead, whose owner may have given his name to the canyon. Around a small grove of cottonwood he had strung steel cable to make a corral. Here and there along the stream in the upper section are fine, shady glades of cottonwood that give the bottom of the valley something of the cultivated aspect of an English deer park after a long spell of unusually dry weather.

But when my friend spoke of Johns Canyon, he was not talking about this domesticated and accessible upper stretch. He meant the part that lies below the ford, which is near a rock-rimmed pool beneath a high sandstone bluff. From there on down, the entire character of the canyon changes. The stream begins to cut deeply into the floor of the broad valley. The valley in turn narrows and becomes steep-sided.

The whole of Johns Canyon is fairly short and small—compared to some other canyons—and drains an area of only 26 square miles, more or less. The canyon stream rises from runoff and seep water, supplemented by a small spring about five miles downstream from the canyon head. But when the weather is very dry in summer, the stream evaporates entirely before it completes its 12-mile course, ending at a drop-off of about 150 feet into the San Juan River. Just before the drop-off, in the lower reaches, the canyon walls rise 1,200 feet above the stream bed, rendering this stretch virtually inaccessible unless you make your way into it from the upper part of the canyon.

One May morning, with photographer David Cavagnaro and his assistant, Dave Nordstrom, I set out to see this private part of the canyon. As we started to hike downstream, sego lilies bobbed all around us like a sea of tethered butterflies; on some blossoms, harvest mites were feed-

HARVEST MITES ON A SEGO LILY

ing. The sego lily (also called mariposa, the Spanish word for butterfly) is almost certainly the most beautiful flower of the canyon country. Most of the year the plant is just a skinny stem and a few leaves no wider than blades of grass. But in the spring, long, candy-striped buds appear first, and then the delicately hued blossoms themselves.

Not far downstream we came to a 150-foot waterfall, a barrier impassable to the livestock that grazes the upper valley. We were able to bypass the obstacle only by working our way along a ledge and then by climbing down a precipitous talus

SANDSTONE RAMPARTS AROUND A POOL IN UPPER JOHNS CANYON

LEAVES OF GAMBEL OAK

slope. Here the rocks along the stream had been sand dunes millions of years before. Now their faces rose up out of the stream at sharp angles, capped six feet above the bank by a horizontal layer of rock. This formation, geologists have decided, was created in a process that began when wind laid the original dunes flat. Then a new layer of sand was laid down atop the old, and both were compressed into sandstone over the millennia by the weight of other material settled atop them but long since scoured away.

The loose sand of present-day dunes along the stream bed was covered with Indian rice grass, whose roots secrete a sticky substance that clings to grains of sand. The root network thus acts as an effective stabilizer for the dunes, protecting them from the eroding effects of wind and water. In times past, the plant's seed,

A PETRIFIED SAND DUNE

finely ground up, served the Indians of the area as a sort of flour. So did the acorns of Gambel oak, a miniature member of the oak family that grows extensively along the stream in the upper part of the canyon.

An Easygoing Snake

Dave Nordstrom was ranging up ahead of us when we heard him call out that he had found a snake. We came hurrying up on the run, just in time to see the reptile's head withdraw from view in a quick blur of yellow and gold. But it hadn't been able to get very far. I found the four-foot creature—a Great Basin gopher snake—hiding beneath the overhanging shelter of a rock nearby.

When frightened, gopher snakes often make a great show of hissing and vibrating their tails, achieving a pretty fair imitation of a rattlesnake. But this gopher made no serious objection when I picked it up and put it into the branches of a dead juniper. After five minutes or so the

RED-SPOTTED TOADS MATING

A SPRIG OF INDIAN RICE GRASS

snake decided that we represented no danger at all and slithered back down to the ground without haste and found the shade of a bush.

Where prey is plentiful, the gopher snake is a prodigious eater of mice—a scientist once found 35 in the stomach of a single snake. But that, I think, must have been in an alfalfa field at haying time. Here in the canyon, where no such numbers of rodents were available, the gopher snake would have to make do with the occasional packrat or small kangaroo rat, like the one whose tracks we saw in the sand nearby.

Back at the stream bank we came across two red-spotted toads mating in shallow water. The male fertilizes the eggs—sometimes totaling thousands—as they emerge from the female. Desert toads like these owe their ability to survive at all as a species to such fecundity, and to the speed with which the young pass through the vulnerable egg and tad-

GOPHER SNAKE IN A DEAD JUNIPER

TRACKS OF A KANGAROO RAT

pole stages. The eggs of the familiar garden toad can take as long as 12 days to hatch; the eggs this female red-spotted toad was producing would hatch in one and a half to three days. In a land where water is often present only for a short time, such haste is necessary.

Right now, with water plentiful after the wet winter, the red-spotted toads were profiting by turning out hatch after hatch: tadpoles in all stages of development swarmed by

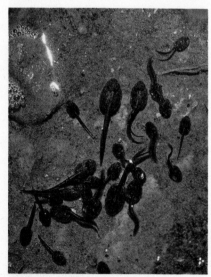

TOAD TADPOLES IN A POOL

the thousands in the shallows and puddles. Plainly, most of those in small, isolated puddles would be unable to find enough food and would not survive. But even so they serve the species: their decomposing bodies enrich the organic matter on the bottom of the pools, on which surviving tads would feed.

Although many of the canyon's animals had already sought shelter

STREAM WATER SPILLING FROM A SANDSTONE OVERHANG

from the day's heat, the lizards were still active. Most of those we saw were northern side-blotched lizards, abundant throughout the canyon country. Every few yards one would go skittering out of our path and pause at a safe distance to watch us go by. Yet even lizards hide from the sun during parts of the day, and we were glad to do the same for a while when we came upon a delicate waterfall with a cool recess beneath it.

A Successful Tree

Through a curtain of water we could look out at the pool below the fall, with a patch of tamarisk growing on its far bank. Tamarisk is an experiment in transplantation that regrettably proved all too successful. The tree, also known as the salt cedar, was introduced to the West from Europe a number of years ago because of the beauty of its pinkish-white blossoms, examples of which now brightened the branches across the pool from us.

Except when it is in bloom, however, I find the tamarisk a sorry, scraggly-looking thing, an appearance nevertheless contradicted by the tree's extreme toughness and adaptability. Tamarisk has succeeded in spreading along many waterways in the southwest, crowding out the native cottonwood, forming all-but-impenetrable thickets and, with its very efficient taproots, lowering the water table wherever it grows.

Much prettier than the tamarisk blooms were the flashes of yellow we were beginning to spot here and there. These were the flowers of ephedra, also called Mormon tea, a

A SIDE-BLOTCHED LIZARD

FLOWERS OF MORMON TEA

A TAMARISK IN BLOOM

leafless plant that spends most of the year as a collection of bare, jointed stalks. (The plant's leaflessness minimizes evaporation, an advantage in an arid climate.) But now bunches of tiny yellow flowers puffed out at each joint. The Mormons made a species of tea from ephedra, which is proof to us that life was tough indeed for the early pioneers. Boiled in water, ephedra tastes just like boiled water to me. But many people say it makes a fine drink, and I won't dispute the point.

Although the walls of the canyon were now very high, they were still set far enough apart so that there was little shade. The only shelter came from overhanging ledges. The plants around us were not growing in profusion. Harsh conditions had caused them to be spread out, and they offered—like the blackbrush in

SPIKES OF PRINCE'S-PLUME

THE MIDDLE CANYON IN AFTERNOON GLARE

yellow bloom near the ledge—little to attract a forager. Blackbrush is browsed by sheep and cattle in the winter, when nothing better is available, but its small leaves and woody stems are nothing that even a camel would walk a mile for.

Deadly Desert Plants

In the struggle for survival, many of the canyon country's plants and animals go several steps beyond the blackbrush. More than being simply unappetizing, they are outright poisonous. We found one of these, a species called Jimson weed, near the canyon wall. Its leaves were uncommonly large—to the inexperienced eye, excellent forage. But the plant, which contains scopolamine and atropine, is deadly. And although a distillate of Jimson weed is used by certain Indian sects in tiny doses to induce visions, in doses only slightly larger the drug is fatal.

Another showy flower along our way was also poisonous: prince's-plume, whose spires of bright yellow blossoms caught the sun on all sides like golden candles. Prince's-plume collects a toxic element called selenium from the surrounding soil. Even so, the canyon-country Indians, after boiling the leaves in several changes of water, used to eat them as a pot herb.

In the shade of an overhang, we came across a creature that is much more of a potential danger to animals and man: a female black-widow spider. Drop for drop, the venom of the black widow is said to be more deadly than that of the rattlesnake. Although the rattler injects

LEAVES OF JIMSON WEED

BRANCHES OF BLACKBRUSH

A BLACK WIDOW

far more poison, the bite of the spider can be equally dangerous. This particular specimen, suspended upside down from a single slender thread in her irregularly patterned web, displayed a perfect example of the familiar scarlet hourglass design on the bottom of the abdomen. In some specimens, however, the hourglass may appear only as a few small spots, or may be absent altogether. Thus the general size—a body about the size of a dime—and the remarkably long pairs of front and rear legs of the adult black widow are more reliable guides to identification. In extensive researches aimed at finding something lovable about the black-widow spider, I have only come up with the fact that, contrary to popular belief, the female does not invariably eat the male after mating. If not hungry, she will have the grace to let him escape.

Farther on down the canyon, the walls closed in slightly around us.

A SPHINX MOTH SIPPING NECTAR

But the time was still midafternoon, with the sun directly enough overhead so that grottoes and overhangs offered the only shade. Such plants as ephedra and prince's-plume can grow in the full sun, but others require shade at least part of the day, and still others thrive in spots where the sun's rays strike directly for only a few minutes each day—or not at all. Around a bend in the canyon, in partial shade, we came on such a community of plants. They comprised a vertical garden, of the sort you sometimes encounter on canyon walls where seeps are present. Often such water, trickling along bedding planes in the rock, is a more dependable source of moisture than the stream running below it.

This small hanging garden, just a few feet above stream level, was principally composed of maidenhair fern and a grass called dropseed. Dropseed often grows in sand, where its sticky roots perform the function of binding dunes together just as those of Indian rice grass do. But maidenhair fern almost never manages to take root on the canyon floor, preferring the well-watered shelter of the seeps in the wall. With mai-

FRONDS OF MAIDENHAIR FERN

DROPSEED GRASS STEMS

A HANGING GARDEN ON THE CANYON WALL

denhair fern are often found such other plants as rockmat and redbud. These favor the same moist conditions of the seeps, which are frequently highly charged with lime salts leached out from the rocks.

A Clumsy Moth

Columbine, a plant with showy white flowers, also grew in the little garden, and we saw a white-lined sphinx moth sipping nectar from one of the gracefully drooping blooms. The sphinx moth is remarkably similar to the hummingbird in general shape, size and function. Both can hover. Both have long tools to reach the nectar deep inside flowers: a hollow retractable tongue for the insect, an extended beak for the bird. Although the insects and the birds took separate evolutionary paths millions upon millions of years ago, the sphinx moth and the hummingbird have developed the same general stubby-bodied configuration.

The hummingbird, nevertheless, seems to me to have solved its structural design problems with somewhat greater efficiency than has the moth. Where the bird uses its beak with deftness and accuracy, the moth gives the impression of being overbalanced when its enormous proboscis is unrolled for action. The sphinx moth feeding before us often missed its mark when it tried to obtain nectar from a new blossom, and then had to back off a few inches for another miniature charge with its cumbersome lance.

At this hour of the afternoon, little but the hungry sphinx moth (which normally feeds at twilight) stirred in the heat. Here and there along the way we had seen mule-deer tracks and, once, we had come across the pad marks of a bobcat. Ant lions waited invisible in the bottoms of the tiny pitfalls they had dug in the sand to ambush ants.

A few lizards scuttled away as we passed; now and again we saw butterflies or grasshoppers. But by and large, life was notable by its absence.

EVENING-PRIMROSES

David Cavagnaro's theory was that the unusually wet winter that had just finished had produced a population dispersal by making the normally arid uplands—as well as the immediate vicinity of the stream —habitable for the moment. This idea was probably correct—though at all times life is relatively scarce in the Canyon Lands.

And yet a few hundred million years ago, as we could see from a close look at the rocks, the country had been heavily populated. All around were fossilized brachiopods, a sort of ancient shellfish. More rarely we came across fossil snails and what seemed to be the fossilized burrows of large marine worms. When I stretched out on the slickrock to take a drink from a clear pool, I saw that its stream-polished surface was made up of millions of tiny shells and other remnants of marine life.

As the heat of the afternoon began to ease off, we came around a last great bend, near the point where the Johns Canyon stream plunges over a couple of waterfalls down to the San Juan River. Now that the sun had begun to hit at an angle, the bottom of the canyon was in the shade. Already the evening-primrose had opened its flowers, and it was time to start heading back.

MUTED COLORS OF LATE AFTERNOON AT THE CANYON'S FINAL BEND

4/ The Mesas

*Above and between the canyons is the slickrock
benchland, that weird world of hills, holes, humps
and hollows where, they say, the wind always
blows and nothing ever grows.* EDWARD ABBEY/SLICKROCK

The word mesa calls to mind a single formation with visible metes and
bounds, a monolith sticking up sheer-sided and flat-topped and finite
from the plain. The image is a true one, though there are mesas so large
as to seem not like features of the land, but like the land itself. Cedar
Mesa in southeastern Utah is of this vastness—more a plateau than a
mesa. Such sweeping uplands may be laced by complexes of canyons,
carved in their trenches far below the surface, the abyssal deeps of the
red rock country.

Some mesas are low and sparsely covered with sagebrush and grasses; some are high and moist enough for juniper and piñon pine; some
are even higher and have enough water to support ponderosa pine. But
all are basically arid, all are wind-swept, all are mainly flat and naked
to the weather. And all are tough places, habitable only by the toughest
of plants and animals.

The great controlling factor in the life of these mesas is water—or
rather its absence. I remember one August afternoon, driving along the
plateau in northern Arizona near Black Mesa. To the northeast a cloud
stood against the sky. Wispy gray sheets hung suspended from its belly
—sheets of rain, but rain that evaporated before it could reach the
parched desert below. Here and there other clouds walked the hori-
zons and some of their trailing gray veils did brush the ground. My car
ran through one of these places just after the rain had passed, while the

air was still full of the superb perfume released and distilled by the rain from the baking land. For a little time longer the air would remain humid and sweet, but very quickly most of the rain water would evaporate back into the arid atmosphere.

Even when heavier rains come to a mesa, they are of little more use than these evanescent showers of summer. The water lies on the tabletop like a thin sheet, moving slowly across the nearly flat land until some slight depression leads the flow to a gully, thence to a wash, then a canyon, and on down a distant river. Moving, the sheet of water lifts up and carries off the dead leaves and twigs from desert plants that might otherwise form an organic base for soil. The water leaves behind only sand, depleted of most of the elements that support plant life —and not even much sand.

Because the loose surface deposits of sand are so dry and porous, rain water will penetrate them to a depth of eight times the amount of immediate precipitation. Replenished by occasional rains, there remains almost forever a zone of moisture where plants can root, with dry sand below and above. But over most mesas in the canyon country, erosion has not left enough sand for such a moisture belt to form. The rain that does fall tends either to run off or to evaporate away.

The result was brought home forcefully not long ago to a friend of mine. A canyon runner for many years, he decided one day to cut short a hike up Grand Gulch about 25 miles from Mexican Hat, Utah, and take a straight compass heading over the height of land back to his truck. The plateau through which Grand Gulch runs at that point is called Pollys Pasture, which hardly conveys an accurate impression.

"I carried all the water I could manage," my friend said, "so I made it out all right. But my God, it's scary country. Up there on top there's no shelter, no shade, no water, no animals. Nothing."

Nothing but the elements. When the heat comes from the sun, it hits these exposed mesa tops hardest. When the winds and the cold come, they do the same. There is virtually no soil, and springs are nearly nonexistent. Yet life claws its way into the rock and hangs on somehow. Plants are spaced widely apart, for each one needs a large area in which to search out what little water and nutriment there is. The plants form little islands between which run avenues of red-brown sand, bare of life. A vulture rides the air currents far overhead, a lizard may scrabble unseen in the dry brush. Nothing else moves.

In Canyonlands National Park, the Colorado River makes two great serpentines at a place called The Loop *(pages 18-19)*, a few miles above the Colorado's confluence with the Green River. At this point, two isolated mesas, once joined by a thin neck of land, rear above the stream. But the connecting razorback ridge has eroded down so far that now each mesa is separate and is protected by its own barricade of cliffs. In aloofness and grandeur and in a sense of barren mystery, few mesas of the Canyon Lands can compare with these two.

"I've been all the way around both of them looking," a ranger in Canyonlands National Park once told me. "There's just no way to get up on top, as far as I can see, and I never heard of anybody else that found a way, either." But there is a way up, despite the ranger: all you have to do is cheat and hire a helicopter.

We—my companion was Hal Mackay, a University of New Mexico botanist—set off from Durango, Colorado. At first our helicopter flew over the San Juan Mountains, with their panorama of upland meadows, ponds, streams, and forests of Gambel oak, aspen, ponderosa pine and Douglas fir. This portion of the San Juans stands at the very edge of the Colorado Plateau, 130 miles from The Loop, and yet the difference between where we were and where we were going was as great as the difference between the Alps and the Sahara.

As we skimmed along, the mountains settled into foothills, the foothills flattened out gradually, the forests gave way to fields of wheat and, later, pinto beans. Below us, next, lay the featureless expanse of the Great Sage Plain, between the La Sal and the Abajo mountains. Then a canyon appeared, and after a while another and another, but seldom enough so that the sight of each was an event. Finally, as our northwesterly course brought us closer to the Colorado River, we realized that we had slipped into the Canyon Lands proper.

On every hand were weird shapes from dream or nightmare, made of bands of white and brick-red sandstone. Here was a long ridge, rising up out of the land like some lumpy scar; there was a bulbous column, or a line of jagged pinnacles standing apart, or a series of scalloped bays gouged from the side of a butte as if by an ice-cream scoop.

From high above, the two mesas we were approaching looked like giant teardrops lying side by side, head to foot. The larger appeared to extend slightly more than a mile on its long axis, the smaller about half of that; each fell off in a vertical plunge of more than 600 feet. The helicopter drew near the larger mesa and we descended toward a landscape that seemed lunar in its magnificent desolation.

"Not likely to find any empty beer cans up here," our pilot said.

Nor even coyote droppings or bobcat tracks. For on these isolated patches of rock, even if a coyote or bobcat—or mountain sheep, mule deer or mountain lion—were able to climb the guardian cliffs, it would be incapable of surviving for very long on such a limited and waterless range. These two mesas can support only small animals and plants, ascetics with modest needs.

It is a curious feeling, to believe that you are the first creature larger than a vulture ever to visit a place. Rationally, you may know you will find a plant and animal community not too different from that on any neighboring mesa of similar altitude. But your emotions and imagination make all things seem possible: it may be that giant sloths, trapped eons ago by the river's meanders, survive up here still; or a camel or a sabre-toothed tiger—both of which actually could have roamed the Canyon Lands as recently as 10,000 to 15,000 years ago.

For a moment it seemed as if the emotional level had been the correct one. The first plant to catch our eye when we stepped out of the helicopter—a prince's-plume—was a monstrosity. Two of its many chest-high stalks ended not in the familiar spires of tiny yellow blossoms, but in hideous, deformed, clublike shapes. They were green, these things, and studded with spikes like a mace. They reminded me of science-fiction horrors, triffids—hostile life forms—from other worlds.

"Could be chromosomal damage from solar radiation," said Hal, the scientist. "Or insects, or chemicals in the soil could have caused it."

Farther away from the helicopter we saw Indian rice grass, yucca, Mormon tea, sage, gumweed, prickly-pear cactus, rabbitbrush, the sour-tasting lemonade bush, bottlebush—all common plants throughout the Canyon Lands. But these, too, were quite different in various ways. The prickly pear that would have been knee-high in a better location barely rose above shoe-top level. Most of the other plants were smaller than normal for their species, and more widely spaced. They seemed more wizened, tougher, somehow more dogged. Biologists speak of something called a tension zone—an area where two ecological zones overlap and where species from both zones struggle to adapt to the borderline conditions. In a sense, aridity, heat and cold had made the top of the mesa into a tension zone for all these plants.

"Horrible soil," said Hal, scratching at the ground. "See, there's virtually no organic matter. But the plants are still hanging in there. Look at this sample down here."

He pointed to a tiny, low growth of the buckwheat family, improbably clinging to life on an inch of wind-blown sand that had settled

The cold glow of a January sunrise, silhouetting the peaks of the mist-shrouded La Sal Mountains, reveals a squirrel's footprints in the snow at Red Sea Flat, Utah. Such heavy snow cover is rare for this area, which averages only six inches a year. But this, with an annual six inches of rainfall, is enough to support shrubs, grasses and scrubby piñon pines, whose nuts are hoarded by squirrels for winter food.

into a depression in the solid rock. But the buckwheat had something special going for it.

"Except for the grasses and the plants that have bulbs, like the mariposa lily," Hal explained, "every plant up here has roots that go deep into cracks. If you pulled up that buckwheat you'd see it had found a crevice to stick its foot into. Up here," he went on, "the number of species is limited. At this altitude in the mountains around Albuquerque, we'd find maybe 150 to 200 different species; here I wouldn't expect to see more than about 30."

In such a dry world as this, plants use three principal means to stay alive. The first is called drought tolerance—the ability to withstand dehydration and still live. Here we could find no examples of drought tolerance—which is by far the rarest of the three survival techniques.

The remaining two are drought avoidance and drought evasion. The former involves the efficient use of wet spells to suck up and store available water in a hurry. In the latter, a plant completes its life cycle hastily during what wet season there is and passes the dry spells as fruit or seed. Every plant around us used one method or the other, or both.

The stunted prickly pear, and a deeply creased barrel cactus not far away, had shallow roots spreading out on all sides to catch water near the surface, and taproots forced into cracks in the rock to pick up whatever water percolated more deeply. When water became plentiful, the presently flat pads of the prickly pear would plump out to provide storage space, and the creases on the little barrel cactus would fill out like accordion pleats.

Other plants that practiced drought evasion were difficult to find at this season; it was a time of drought and they were, by definition, evading. But when the snows melted next spring and when the summer thunderstorms came again—the two great blossoming seasons in the Canyon Lands—the plants would return to vigorous life. The seeds of the yellow bee-plant, the Russian thistle, the scorpionweed, the lupine, the sunflower, all would stir and sprout, giving the dun landscape their accents of bright color. Many of the drought-evading plants would flower, fruit and go to seed in just a few short weeks, and would be seen no more for the rest of the year. Moreover, these desert flora are obliged to be not only quick but also highly efficient and productive engines. They must convert the sun's energy and the soil's scant nourishment into astronomical numbers of seeds. A single Russian thistle, for example, may produce 85,000 seeds—a necessary excess for survival since most of the seeds will fall upon stony places or be eaten by animals.

Walking a bit farther, we saw a four-winged saltbush, hung with pale-yellow fruits that looked like hundreds of tiny, four-sided (or four-winged) Oriental lanterns. The saltbush survives by drilling for water with a thick taproot that can far outweigh the visible portion of the plant. Scientists have dug out these roots to the incredible length of 200 feet. This extensive underground plumbing supplies nourishment to more than just the plant; all along the saltbush's branches were little growths like cotton puffs, each sheltering an insect larva about a quarter of an inch long. These would someday transform themselves into tiny scale insects called *Eriococcus tinsleyi,* which would in their turn be parasitized by the larvae of an even smaller, one-millimeter-long wasp named *Aphycus howardi.*

On the other side of a slight rise, Hal came across a single-leaf ash. The tree was no more than a bush, but it was far from young. "It could well be 50 years old," Hal said. "Plants like this are right on the edge of survival. They have to fight years for every inch of growth." The tips of the twigs on the miniature ash were ringed with small ridges, each one representing the growth and death of a single leaf. We counted 15 such ridges in the space of a single inch.

Approaching a slight depression in the mesa top we discovered what was surely the oldest and the biggest living thing there. It was a desert hackberry, a tree that elsewhere might have stood nearly 30 feet high, and normally is found only in areas with more water. This one, huddled in a bone-dry wash, stood about eight feet tall, with a trunk measuring about 10 inches across at the base. Yet, not only was it fully adult, but very old—Hal estimated its age at between 250 and 500 years. Since the time of Columbus, perhaps, this tree had been fighting off all challenges to its existence. And each year, as well as merely surviving, it managed to grow a tiny bit.

The hackberry owed its great age to the fact that it had been born with what amounted to a silver spoon in its mouth. Centuries ago, a seed had rooted in a place that just happened to be the most likely on the mesa: at the lower end of a long, shallow dip in the ground. Water from a considerable area could thus be funneled down to the hackberry. Moreover, just behind the tree was a small cave, perhaps 15 feet deep and high enough for a man to stand erect. Here snow could collect during the winter and linger, shielded from the sun well into spring, slowly releasing meltwater to the tree's roots. Furthermore, the tree stood on a south-facing slope, so that in late autumn and early spring it would catch a little extra warmth. The hackberry had built on each of these tiny advantages to outlast and outgrow all competition, until it had become the boss plant of the mesa top.

Working our way up the drainage from the tree, we came upon a spot on the ground scattered with buff and gray mourning-dove feathers. A nest, empty now, sat on the rock wall nearby. A hawk or an eagle must have struck the dove and carried it off; no trace of the bird was left except the few feathers, and aerial predators would be the only ones able to reach so inaccessible a bastion.

Just down the rock face, in a crack, was a different sort of nest, a seemingly random construction of webbing that formed a tunnel leading several inches into the crevice. Outside, two egg cases the size and color of small mothballs hung in the webbing. When I poked with a stick at the cases, a dark shape moved in the tunnel. I had been careful to pick a long stick, because I knew what was moving in the shadows. Later I fished her out with the stick—a large spider, glistening black, with an elegant scarlet hourglass on her abdomen. A black widow.

Above the spider's lair near the top of the gently sloping wash, were patches of reddish, sandy soil whose surface crunched underfoot with a brittle noise. The sand was broken up and eroded into tiny battlements and spires, uncannily like those we had just flown over on our way into Canyonlands National Park. All the park's weird, tortured, beautiful rock forms seemed to be reproduced right at our feet, and the processes that had produced this tiny landscape were not too different from those that had formed the great one all around us. The sand had been of fine, uniform consistency and density to begin with, but capillary action had drawn rain water up to the surface, carrying various minerals in solution. When the rain water evaporated, it left these minerals as a hard crust on the surface—like a more resistant stratum of rock overlying softer ones. But the tough, new surface had not been precisely uniform, any more than is a stratum of rock. Parts of the crust were thinner than the rest and had collapsed here and there; other parts were weakened by holes or cracks. Rain and wind had enlarged these openings and had then worn away the soft, sandy underpinnings to create this strange miniature landscape.

Here and there among the inch-high ramparts, tiny buff-colored curlicues came out of the ground. These were the seed tails of stipa grass, which has worked out an extraordinary adaptation to its harsh environment. The seeds of stipa grass (sometimes called needle-and-

Barren and exposed, the weathered sandstone at the summit of The Loop supports a bare scattering of drought-resistant blackbrush.

thread grass) do not leave the planting process to chance. They actually drill their way into the soil—as they had done here.

The seed itself is enclosed in a case, pointed at one end and so hard that it can injure the mouths of horses. Botanists call this case a callus; it is the tip of the stipa grass's drill. The shaft of the drill is the stem (called the awn) to which the seed is attached. When the wind carries the whole assembly to the ground, the heavy, sharp seed part strikes first, like the weighted business end of a dart. The shaft, which is twisted in its natural growth as tightly as the rubber band powering a model plane, stays wound up until moisture hits it. Then it unwinds. When the shaft later dries, it twists itself up again. As wet and dry periods succeed one another, the shaft screws itself into the ground, pushing the seed down in front of it. In our miniature landscape, we found one seed that had drilled itself a full two inches into the soil.

Out of the shelter of the little draw, on a patch of high ground, lay a huge slab of red sandstone. Littering the crevice under it was a pack rat's debris: seed husks, leaves, odd twigs and pebbles. The pack rat's midden reminded me of another small rodent of the desert—the kangaroo rat. An animal with no sweat glands, the kangaroo rat drinks no water. He acquires all the moisture he needs from the food he eats, and he can cool off by breathing, since the cooler interior temperature of his nasal passages allows some moisture to be retained by condensation. But for all the help he gets from these adaptations, the kangaroo rat's ability to survive the desert's heat and dryness is a triumph of technique rather than biological specialization.

Consider what happens to a man, a somewhat larger warm-blooded animal, in such places as this mesa top. He must keep his body temperature within just a few degrees of 98.6°F., or die. If the air temperature is much above this, he must cool himself principally by evaporating off sweat. The source of the perspiration is the water in his blood, which desert sunlight will tap at a rate of about a quart an hour. By the time two gallons are gone—about 10 per cent of body weight —the man becomes delirious, deaf and, rather more happily, insensitive to pain. When another 2 per cent is gone, the blood thickens like boiled-down gravy so that the heart can no longer circulate it effectively. Thus the blood is unable to carry off heat from the inner organs to be dispersed from the skin, and the man very rapidly dies.

If our man knows what he is doing, he can drag out the process somewhat. Saharan tribesmen, for example, battle dehydration by bundling

Thread-thin tails of stipa-grass seeds thrust above a miniature landscape of eroded sand turrets near the top of The Loop. The grass's seeds are buried in the soil; their tails, known as awns, curl in spirals during dry times and uncurl in the presence of moisture. This alternating action literally drills the seeds into the ground. Shown here on a damp day, most of the tails are unwound, though one of them (left center) still holds some of its curl.

up, which seems illogical to one who thinks of clothing as something to keep the body's heat from escaping. But the Saharan Tuareg's problem is the reverse: keeping the coolness and moisture of the body from escaping to the hotter and drier oven outside. Even his facecloth serves a purpose: the moisture in his breath, which would otherwise be lost, is trapped and can be reinhaled. Lacking such protection, a man can still cling to life for a while—by finding shelter in a cave, moving only at night, and seeking out such modest sources of moisture as cactus flesh, yucca blossoms and wild rhubarb. But stranded on this mesa top, he will eventually die nonetheless. His water needs are too great.

However, using much the same techniques, a small rodent such as the kangaroo rat can drag out the process for a whole lifetime. If he burrows down—as he does—a mere 20 inches below ground level, he experiences hardly any variation between day and night temperatures. At that depth, a temperate mean prevails. And at twice that depth the year-round temperature would remain nearly constant, though the air outside might be zero in winter and more than 100° in summer.

The kangaroo rat's hole performs much the same office as the Tuareg's robes. The animal's own moisture, as well as any from rainfall that might be trapped below the surface, surrounds him in a pocket of underground humidity. Since there is no wind to carry it off, and no excessive heat, loss of body moisture is reduced to a minimum.

Yet the loss that does occur is somewhat offset by the rat's life techniques. Though the seeds that make up the bulk of the kangaroo rat's diet—and also make up his water supply—are virtually dry when he finds them, the rat, instead of eating them right away, carries them back underground in his cheek pouches for storage. At the end of several weeks in the humidity of the burrow, the seeds will have absorbed from the air as much as a third of their weight in water.

When the kangaroo rat selects seeds to store (working at night to avoid the desiccating heat), he is aided by a sense of smell keen enough to detect which seeds are rich in sugars and starches, and which are rich in proteins. The first two kinds he packs into his cheek pouches, the last he tends to pass up. This is because all animals, when they break down proteins in their diet, produce nitrogenous by-products. These must then be flushed out of the system by urination, which means the loss of water. In any environment where water is scarce, it becomes wasteful for an animal to consume any more protein than is strictly necessary to maintain health. And here, on the mesa top, to be wasteful of water is to be dead.

Framed between a stand of rough grasses (foreground) and a low ridge, a field of stipa grass waves in a May wind. Luxuriant in the

aftermath of an unusually wet winter, the stipa glistens with sunlight reflected from many tiny silver hairs atop each seed-bearing stem.

To be abroad in the heat of the day is also, for many animals, to be dead. I lived in Morocco for a while; there I once found a scorpion under a stone behind my house, and put it in a jar to show my children later on. Without thinking I set the jar in the direct sunlight, and when I returned 10 minutes later the scorpion was dead. In the full heat of the desert day, only the lizards risk the sun. One sat on the edge of the pack rat's sandstone slab, looking us over. It was a collared lizard, a large and handsome species. Unlike the rodents, the lizard has no fear of the sun. But then, his problem is the opposite from that of the warm-blooded rodent. The rodents' machinery aims at keeping internal temperature stable within a very narrow range, despite wide variations in the outside temperature; the cold-blooded reptile depends on that outside temperature to keep an internal temperature somewhere within a relatively wide operating range.

During a brief March snowstorm outside Aztec, New Mexico, I came across a small lizard too chilled to move when picked up, but still some distance from the absolute lower limits of his tolerance for cold. To test these limits, zoologists have immersed lizards in ice water until the animals became not only motionless, but so stiff that their bodies could not be bent with the hands. Although their metabolism had slowed down greatly, the lizards quickly recovered whem warmed. But in the canyon country, cold-blooded animals must also cope with temperatures, in direct sunlight and at the surface of the ground, that regularly reach as high as 158°. Heat of such severity will kill even lizards in very short order.

Thus the collared lizard on the rock before us had spent most of his life in the circumstances of a man with a painfully hot fire before him and the arctic chill behind. A man would keep moving around, and so do lizards. But they do other, more ingenious things too.

Some (the collared lizard is one) dissipate body heat by panting as does the kangaroo rat. Some climb into bushes to avoid the fierce heat of the desert floor (only a few inches above the ground, the temperature can be many degrees cooler). Some escape the heat just at ground level by rising on their toes at intervals. Some lizards have been seen to stop suddenly when running across wide, hot stretches, and to rub their bodies from side to side so that their bellies come in contact with the cooler layer of sand just beneath the surface. And all, of course, seek out the shade when the sun gets too hot.

When the temperature is so cold that a lizard becomes too sluggish to hunt or to escape from predators, he reverses his survival techniques.

For example, a lizard venturing out in the cool of the morning will seek the sun and arrange his body perpendicular to its rays so as to warm up more quickly. The lizard's skin color, too, may darken so that more heat will be absorbed into his body. (Specimens killed by cold are notably darker than they would normally be.) By using this combination of stratagems, lizards soak up heat from the sun so effectively that herpetologists have recorded an 86° body temperature in a basking lizard even though nearby shade temperatures were below freezing.

The collared lizard before us showed no particular alarm at the near presence of two impossibly huge giants. When we approached closer than a yard, he only scampered out of range. I made a quick move, to try to startle him into flight, but all he did was disappear almost casually into a crack under the rock. I had hoped that he would get up on his hind legs like a tiny *Tyrannosaurus rex* and then run for his life in this erect position. The collared lizard sometimes does this, although I have never seen it. People who have say that the little animal seems to be leaning backward as he scoots along, looking like the Roadrunner in the movie cartoons.

After the lizard had disappeared, we wandered a little longer on the mesa top, with the idea of getting to the edge. But the small draw that we thought might lead us there turned out to be impassable and we hadn't enough time to look for another way.

Instead, I went back on the ledge above the ancient hackberry tree to see if from there I could get a broader sense of the isolation, the essential apartness, of our mesa. But I saw something else entirely. In the distance on all sides were dozens of other mesa tops. Some were a little higher than ours and some a little lower, but most were about the same. Here on the top of a vast tower, I lost the sense that it was the top at all. Our altitude seemed to be the proper and normal one, and we to be visitors on one of an endless chain of deserted islands, between which the sea had unaccountably disappeared.

5/ The Rivers

Turning around, we can look through the cleft through which we came and see the river with towering walls beyond. What a chamber for a resting-place is this! hewn from the solid rock, the heavens for a ceiling, cascade fountains within.

JOHN WESLEY POWELL/ *THE EXPLORATION OF THE COLORADO RIVER AND ITS CANYONS*

Looking down from the brim of some awesome canyon at the tiny ribbon of muddy water far below, one is seized with the normal wonder: How could such an apparently inconsequential stream have carved all that? The answer is that it couldn't have—all by itself—and it didn't. If it had, the canyon would be a slit trench a thousand feet deep and hardly wider than the stream itself; plainly a stream can cut only where it flows. The rivers are not just the carvers of the canyons; just as important, they are the liquid conveyor belts that carry off the debris of erosion caused elsewhere and otherwise.

And the real wonder of these rivers is not that they cut so deeply (although they do) but that they carry so much—especially since their replenishing supply of water is so spare. Even the principal watercourse of the Colorado Plateau, the Colorado River itself, does not appear among the top 24 rivers in America as ranked by the U.S. Geological Survey in order of average water discharge. The region's other rivers —among them the Green, the San Juan, the Paria, the Escalante, the Dirty Devil—are even smaller. For example, for a great deal of its length the San Juan is not much wider—around 150 feet—than the Housatonic River is where it flows past my house in the village of West Cornwall, Connecticut. And yet each year the San Juan carries more than 34 million tons of sand, mud and gravel past the similarly sized village of

Mexican Hat, Utah. During one extraordinary day of flood, October 14, 1941, the river transported an estimated 12 million tons of sediment past Mexican Hat. (By way of contrast, during a 10-day flood in 1965, the Mississippi River carried only 340,000 tons of sediment past Saint Paul, Minnesota.) And when the sediment washed from the San Juan, together with that of the Colorado, settles out as it gradually does on the floor of Lake Powell, each year it covers the equivalent of 80,000 acres a foot deep in sand and silt.

In addition to the sediments, the rivers transport a huge but unmeasurable bed-load—boulders, cobbles and pebbles pushed along the channel's bottom by the current. When the water is high, you can stand on the banks of the San Juan and hear the boulders thunk against each other as they roll downriver, invisible beneath the brown tide.

At floodtimes like these, the rivers of the canyon country become something between a liquid and a solid, almost like a wet avalanche. Even in normal times, when the sediment load may be only 1 or 2 per cent (as contrasted to a flood high of 75 per cent), the dividing line between land and water often seems blurred. The watery sand of the bottom and the sandy water of the current form a single viscous substance in continual flux. On the river bottom, underwater sand dunes are constantly building, being torn down, and rebuilding. If the river is shallow enough, water flowing over the dunes forms waves. These are called sand waves, although they are no more sandy than the rest of the current. Their name comes from the submerged eminences that cause them. Going down the San Juan River recently, my wife and I were both dumped out of our raft by sand waves; fortunately we were attached to the raft by ropes around our waists.

Though such misadventures, and the natural phenomena that cause them, can occur during any part of the wet spring season, the greatest danger occurs when occasional violent thunderstorms hit and rain flushes away loose silt and sand, sending them hurtling down the stream beds with enormous suddenness and savagery. In the grip of a flash flood, a stream may suddenly rise 10 feet or more within an hour—then retreat to a trickle in the next hour as the waters sink into the porous gravel of the stream bed.

The violence of wind and rain that come with such storms strikes not only within the canyons but all across the broad plateau surrounding a river, and the fantastic shapes of the rock all through this country suggest that they have been carved by wind and rain, working directly on the rock like a sculptor's chisel. Indeed, the almost universal as-

Its contours embellished by a dusting of early snow, a double curve of San Juan Canyon, called the Goosenecks, stands as a monument

to the time, 30 million years ago, when the ancestral river first meandered across a plain, eventually slashing this spectacular abyss.

sumption is that these agents are the principal forces acting to pry loose sediment and gravel from the land. But the fact is that the intricately sculpted grottoes, natural arches and ledges of the canyon country are formed mainly as the result of ground water seeping through pores and cracks in the body of the rock, dissolving its matrix until the constituent sand particles trickle away. The winds and the runoff of rain then act largely as transporters—down to the river—of material already eroded free of the rock mass, not only by ground water but by frost. Thus the rivers respond to a complex of physical influences far beyond their banks and even their canyon rims.

Though tourists ride the high water of the Colorado and San Juan on rafting trips, they rarely get a true feeling for the ultimate work of these rivers. Fewer people ever go down into the river bottoms, to the bowels of the canyons, especially the smaller ones.

One spring not long ago I set out with a friend to walk the bed of a small river—the Paria, which joins the Colorado at Lee's Ferry not far below Glen Canyon Dam, at the foot of Lake Powell. We covered the last portion of the Paria's course, between Route 98 in Utah and Lee's Ferry, which is in Arizona. The straight-line distance between the two points is just over 20 miles, but the river lays a gash as jagged as lightning across the face of the Paria Plateau, so that the distance by foot is more like 40 miles.

At the end of May the sun was hot. In its glare the Paria River was pale, the color of coffee, heavy on the cream. In most places, the Paria River, seldom more than knee-deep and 10 or 15 yards wide, would more likely be called Paria Creek. But in country so arid, if you can't step across from bank to bank and if water runs most of the time, then it qualifies as a river.

As we got ready to set off down the river, nothing else moved on in the noon heat. And then suddenly a hideous squawking broke out overhead. Backlighted against the sun, a raven flapped and lumbered gracelessly along while a tiny bird rode its shoulders, pecking. The little bird, a sparrow, was cheeping in squeaky fury, now and then leaving its privileged sanctuary on the raven's shoulders to give the bigger bird a quick peck from the flank or the rear. It was plain that the raven, helpless to do more than loose harsh yawps of outrage, was in full retreat from an incautious venture into the tiny bird's territory. But the sparrow would not allow the raven peace with honor.

Hal Mackay, the University of New Mexico botanist, and I moved on down the river after the raven. At the beginning, the canyon walls were a light gray, nearly white in the bright sun. They showed the wavy laminations in which they had been laid down, eons ago, as sand dunes. The Canyon Lands are full of such petrified sand dunes. Where erosion has shaped the ancient dune so as to lay bare its original surface instead of slicing through it, the fossilized tracks of small vertebrates are often found. For many years geologists puzzled over the fact that these stone tracks virtually always ran uphill, almost never down. But the mystery was solved some 30 years ago, when a scientist noticed that present-day lizards leave beautifully clear tracks on their way up a sand dune; yet the same animal's tracks on the downward slope are usually obscured by the reptile's own sliding and by the shifting of the sand under its feet.

The bottom of the canyon was still wide enough so that the sun could reach it. And yet there were few tamarisks, no cottonwoods, little growth more substantial than reeds and grasses; the canyon bottom was flat, with no terraces where larger growth would be sheltered from flash floods. Underfoot there was only fine sand to slow us down or mud to suck at our steps and stick heavily to our boots. The mud was greenish or grayish or blackish brown. In some places it had dried, and the thinnest muck, on the surface, had cracked into little squares curling upward at the edges like old paint in the sun. Neither of us carried a watch and so we did not know how long it would be until the sun's heat let up. Time in the canyons is linear anyway, a thing to be measured in paces and miles rather than in hours.

After a certain distance the canyon narrowed. As the walls closed in we had to cross and recross the shallow river to reach whichever bank was not a sheer rock wall coming up straight out of the water. Here and there the footing was firm, but mostly it was gluey. The sand in canyon-country rivers may have been several times compressed into rock, weathered out again as sand, transported down streams or by the wind, until each individual grain was nearly as round as a tiny marble. In a stream such as the Paria, sediment in the current is automatically graded by weight as it goes along, with the heavier fragments falling out first and the lighter stuff being carried farther downstream. The result is that many sandbanks found along Canyon Lands rivers, including the Paria, are composed of tiny spheres that have been naturally sorted to similar size. Infused and lubricated by water to the consistency of very thick soup, these spheres of sand form something that is not quite land and not quite stream: quicksand.

The frightening stuff was here and there all along the banks, and often sucked at our feet in midstream as we waded across. The trick is to keep moving. When the surface is exposed to the air, quicksand, with a skin of surface tension on the top, is like cooled chocolate pudding. Move briskly enough and you can pass over the quaking skin without breaking it. Or you can play on it as on a trampoline, shifting your weight from one foot to the other while the whole mass shudders rhythmically back and forth under you. But if you play this way long enough, the skin will finally split, showing the darker, wetter jelly below. Then it is best to move rapidly, and find another game to play. Not that there is any real danger, of course. They say you always hit solid bottom before you sink out of sight. That's what they say.

At one point Hal went on ahead of me while I sat down to change my wet socks and rest. When I caught up, he was washing mud off his clothes and laying them out on a rock. "You're too late now," he said, "but I could have used you ten minutes ago."

He explained that he had slowed down just long enough while crossing the stream for his feet to get stuck in quicksand. His first instinctive moves to jerk his feet free only got him in deeper. Quicksand works like the kind of auto seatbelts that yield to a slow pull but lock tight against a fast tug. Yanking one foot thus has the effect of leaving it just as solidly caught as before, while driving the other foot farther in. Before long, Hal had sunk in almost to his knees. But by moving very slowly he was eventually able to ease himself free of the sand's grip and onto solid footing.

"I would have been along soon, anyway," I said. "You could have breathed through a reed while you were waiting."

Hal muttered something I didn't quite catch.

By now the canyon walls had changed from light gray to red sandstone, which went up nearly straight for hundreds of feet from both sides of the little river. Though we could see a slice of the bright blue sky far above, we were walking in open shade. In a canyon this steep, unless the sun is straight overhead, its direct rays cannot reach the canyon bottom. Twice we came across small pieces of red sandstone fallen down from the walls, each rock surrounded by a little aureole of brick-colored fragments. The sight told us why the walls above rose so steep and so close; their sides are hard enough so that they resist such major erosional effects as landslides or creeps, yet when fragments do break away they are soft enough to come partially apart under the impact of hitting the canyon floor, producing pieces of an easy size for the waters to carry away. Thus the stream bed is kept clear and the river can deepen its course straight down into the rocks that lie below. And, once carved in rock of such consistency, a canyon tends to maintain its steep-walled profile as long as the river retains enough impetus—largely from the grade of its descent—to keep on cutting straight down.

At the bottom of the canyon, almost nothing lived. Around us were the mud and the water, the sand and the rock walls, and very little else. Once we came across a cluster of bats asleep in a grotto, and once a lone seedling coming out of the gray-green mud, a wild sunflower.

"See how it's growing tall and skinny?" Hal remarked. "Like a plant in a closet, reaching up for light that isn't there." The seedling had no chance. The flash floods of July and August would destroy any vegetation that managed to get a foothold here in this sunless slit. Not even the tamarisk, an extraordinarily tough tree, could long survive here. Neither could we, if an early thunderstorm upstream sent a wall of brown water boiling down on us before we could get out of this particular five-mile stretch of vertical gorge.

I saw a sphinx moth belly up, stuck dead in the mud along the stream. Later there was another and another, until we had seen a half dozen or more. Each was on its back, with no clue as to what had brought it down into the dimness of the canyon floor to die. The sphinx moth is as big as a small hummingbird. The moth's larva, a large, bright green caterpillar, eats the foliage of the hairy evening-primrose; the adult moth closes the circle of interdependency between primrose and insect by pollinating the plant's night-blooming flower. But down here in the somber gorge there were no primroses to attract the sphinx moths, nor any flowers at all.

Farther on, I came across a dead sphinx moth that had left a story behind. Although the insect's legs were stuck in the mud, it had died hopefully, body angling out of the muck and toward the sky, striving for upward mobility. All around the dead moth were feathery patterns where its wings had left marks in the goo during the struggle to escape. I guessed that, during similar struggles, the other dead moths upstream had gone through a sort of desperate gymnastic routine: each had first mired one wing and then been flipped over on its back by the powerful beating of the other. But none of this explained why the moths had got caught in the first place. Much later I came across a possible clue in an entomology text that told of a species of sphinx moth known to ingest 173 per cent of its own weight, becoming too heavy to take off. Perhaps, then, the insects had come down into the canyon to tank up on water, had overdone and had wandered around earthbound till they got caught in the mud.

We were a little afraid, ourselves, that we would not get out of the mud before nightfall. There was no way to get really lost as long as we followed the river, but we had no way of estimating how much longer we had to go before coming out of this twisting gorge that formed some five miles of the river's course. However, we knew that the canyon widened a little way past the river's junction with Kaibab Gulch; there we hoped to find proper riverbanks, with places to sleep out of the mud.

Toward dark, we finally came to the junction, which is on the Utah-Arizona border. At first Kaibab Gulch seemed like more of the same: stone walls rising up sheer out of mud and quicksand. A trickle of turbid water wandered through the muck, a current so sluggish as to be barely discernible. We headed up the tributary canyon nonetheless, hoping to find some water that had been standing long enough for the mud to have settled out of it; through the heat of the afternoon, we had been drinking the brown river water. We were sick of the mud in the water, sick of everything.

And then we went around the canyon's first bend.

There before us was a scene so unexpected and so beautiful as to stop us, silent, for a moment. Over the ages, the stream had laid down sandbanks on either side, terracing up to 25 feet or so on one side, perhaps 50 feet on the other. On the first level of the terracing, chest-high along the stream, lay a bank of grass with wild lettuce and wild mustard scattered here and there. Above grew Gambel oak and redbud trees, with an understory of grasses and small shrubs to make up the floor of the miniature forest. Among the larger trees we also could see box elders, small members of the maple family. The whole bend in the canyon made a chamber not more than 100 feet long and about as wide, hidden deep within the earth. The secret room was still and silent, fixed in dim, greenish light like some submarine cavern.

We left after a while, having filled our canteens with clear water from the still pools and picked wild greens for our supper. As long as the light was still holding, we wanted to make what distance we could. Half a mile down the Paria we found a sandbank not nearly so lovely, but suitable enough for a campsite. Two ravens were working the canyon rim far above us in search of small prey. The pair showed none of the lumbering clumsiness of the raven we had earlier seen harried by the sparrow; here they soared on the updrafts with wings almost mo-

A living wedge in the sandstone, a piñon pine takes nourishment from the January snowpack. The piñon's seed, probably carried here on the wind, settled in a tiny rock fracture and sent out roots in search of water. As the trunk grew, it pried the rock apart at the surface. Below ground, a taproot, which may reach as deep as 15 feet, has enlarged the crevice—thus contributing to the erosive process that breaks up the established rock strata and delivers their fragments to the rivers as sediment to be washed away.

tionless, in arcs as fluid and graceful as a hawk's. It is a nightly sight in the canyon country, ravens over the riverside cliffs. At dark they give up the patrol and retire to their holes high on the rock face. Ornithologists who have climbed up to inspect ravens' nests have been less than enthralled by the experience. The raven not only eats carrion, but also uses it for building material. Thus a raven's nest is often lined with such stinking stuff as hair from dead cows and wool plucked from the carcasses of sheep. The result can be noisome, though it doesn't seem to bother the hawks and owls that frequently move in after ravens have abandoned a nest.

The harsh squawks of the ravens stopped as the sun set and in their place we heard the low, sad, clear tones of doves somewhere near us along the river. Every day at dawn and at dusk the doves of the desert fly to water to drink and to bathe. So dependable are the doves that experienced desert travelers will note the direction of their twice-daily flights, so as to know where to look for water in case of need.

By the time the light had died out entirely, my driftwood fire was going well and I began cooking up a pot of rice, dried soup mix, wild lettuce, wild mustard and the spring growth of baby tumbleweed just coming up all over the sandbank. A simple recipe, perhaps, but one that demands full time and attention from the chef. The pot must be hung from a green tamarisk branch, cut and stuck in the sand, firmly propped in place with chunks of sandstone. The heat must be regulated by moving the pot in or out of the fire. The stuff must be occasionally sniffed and stirred to keep the rice from sticking to the pot.

I busied myself doing all these things and smelling the juniper smoke, listening to the calling of the doves, wondering if a rock would come loose from the canyon wall during the night and fall on my head as I slept. I watched the bobbing of the bubbling pot; pretty soon the mess was ready for eating, and soon after that I was ready for sleep.

When we set out again the next morning, two swallows were doing aerobatics over the surface of the water, strafing and dive-bombing each other. The red walls of the canyon were still very high and close so that direct sunlight could reach us only here and there along the way. The river's bed was so tortuous that often there were only a few score yards from one bend to the next. Instead of cutting a course straight downhill across the plateau, the Paria had followed fissures and rock faults that often led the river off on sharp-angled detours. It was like going through a series of long, narrow rooms.

Its base pared down by flash floods, a boulder leans against the canyon wall above the summer trickle of the Paria River.

Around one of these sudden bends, we came on a spot where the sunshine lit a grotto of brilliant green, a splash of deep emerald across the baked red rocks. Ground water had worked along an impermeable layer of rock until it could escape at last here in the canyon wall. Cool water fell through the air in little streams inside the grotto, which was not much higher or wider than a man and perhaps two feet deep. Outside in the sunlight, water seeped in sheets from a seam in the rocks a foot or so above our heads. Masses of maidenhair fern hung below, all along the glistening cliff face. Tiny yellow monkey flowers grew among the maidenhair. Helleborine, an uncommon yellow, brown and pink orchid, was just coming into bloom, and a shoal of small flowers ran along the vertical wall.

The water dripping inside the grotto and seeping down the cliff face outside combined to make a tiny, clear stream that ran along the canyon wall for a bit before it disappeared into the mud of the Paria River. In the clear water we were able to see our first fish of the trip, a school of young silver minnows not an inch long. They were flipping and wriggling in frenzy, apparently trapped on a shoal too shallow even for their minuscule draft. But it turned out that they were just playing at being stranded, and an instant later, all at once, they flipped back into deeper water with no difficulty.

As we went on downstream, we came across more and more of these little oases. It is strange to talk of oases on a riverbank, but this is what they amount to. In the canyon country, riverbanks are not much more accommodating to plant life than is the desert itself. Since very little of the river's water spreads out from its course, large perennial plants must survive by sending roots out to the stream or down to the water table below. Cottonwoods and tamarisks do both. As well as reaching out with horizontal roots toward the stream bed, they send strong taproots straight down.

Such taproots enable plants to survive droughts during which the nearby stream may completely disappear; plants smaller and less well equipped may simply die. In the few places where lush and permanent vegetation appears along the course of a river in canyon country, it is likely to be watered not by the undependable river, but by a reliable seep in the canyon walls. These seeps also nourish the hanging gardens of the Canyon Lands; the river's only role in creating such small, vertical oases is to cut deeply enough through rock to expose the strata along which ground water moves.

Rocks and driftwood scoured by the churning sediment of floodwaters lie in Cataract Canyon on the bank of the Colorado. At the height of the spring floods, the river sweeps along everything in its path—boulders, trees, and any shrubs growing in the canyon bottom. As the torrent subsides, heavier rocks settle out first, subsequently trapping lighter flotsam. This pile of debris, deposited above the normal level of the river, will remain until another flood breaks it up.

Soon the canyon walls began to open up so that there were spaces for substantial sandy terraces at points where the river formed bends. The first of the terraces we saw supported only box elders, not much more than shrubs. Subsequently we began to see Gambel oak and healthy stands of cottonwood, whose little white seed puffs lay thickly strewn on many of the terraces. The tiny, bright fibers of which the tree's cotton is made serve as combination wings and parachutes to carry the millions of seeds on the winds. The fuzz is almost pure cellulose, so inflammable that it flashes away to nothing in a tiny, soundless explosion if you touch a match to it.

By late afternoon, when we began to think about a campsite, the canyon walls had closed in on us again. Thunder began. From upstream? Downstream? From our left or from our right? With the sound thrown around by canyon walls, the thunder's true origin was hard to establish. Moreover, a jet would pass above now and then, out of sight and sounding very much like thunder. Which noise meant a TWA dinner flight, and which a flash flood that would wash us down to the Colorado River? The sounds made by God and by man have got all mixed up, so that a latter-day Joan of Arc might well write off her voices as some electronic malfunction.

Presently the thunder moved away from wherever it had been, but we were careful nonetheless to pick a campsite on a sandy terrace at a safe height above the river. Once we had dumped our gear, I went exploring along a nearby stream. Though hardly more than a seep running down a gully into the Paria, it held enough water to support a stand of bushes and chest-high reeds, below which were stiff stalks of scouring rush that rattled and squeaked underfoot. The jointed stems of the plant contain silica, the same material that makes up quartz. Thus the hollow stems are so abrasive that they can be used, as the plant's name suggests, to scour pots.

Above the clacking of the scouring rush as I pushed through it came the unmistakable sound of a snake: a smooth, continuous rustle quite unlike the noise of any other small animal. The reptile, a harmless striped whip snake, made a sudden dash through the clump of reeds and then stopped dead under a bush. But by moving very slowly I was able to avoid startling the snake and get within range. When I grabbed it, the animal struck out at me once, missing, then settled down directly and seemed perfectly calm in my hands.

The snake was four feet long but no thicker than a man's finger, with a pale yellow belly that turned to pink under the tail. Yellow stripes ran the length of its body on either side. Like all its cousins—the racers, the coachwhips, and other types of whip snakes—the striped whip snake lives a good deal in the open, by daylight, and counts on its speed to escape enemies. This speed seems impressive to the eye, but is largely illusory; scientists clocking a Colorado Desert whip snake found that, at full tilt, it was moving at only 3.6 miles per hour. My own theory is that stripes make a snake seem to be moving faster than it is. More than once my eyes have become so fixed on the stripes of a flee-ing snake that its rapidly flowing body mesmerized me and my grab closed on the air. (Or, in one unhappy instance, on a cactus.)

After bringing the snake to show Hal, I set it down on the sand near a tangle of brush. Instead of making immediately for cover, the animal stayed where it was, apparently content and unafraid. I have often no-ticed this behavior in newly freed snakes and don't know what causes it. Surely it is not the result of an exercise of pure reason in which the snake concludes that, since the monster has been harmless so far, it is no longer a threat. Considerable experience with snakes has convinced me that they are very limited in their powers of reasoning, lying in that respect somewhere between an Air Force general and a White House special assistant. Whatever the cause, though, the snake lay there until I poked at it with my finger, and only then did it disappear quickly in-to the brush pile.

Before we had gone far the next morning, the canyon walls drew apart, so that what had been a giant slit just a few miles north now became a wide valley, rising slowly up from the river until the slope ended in pre-cipitous walls. The river went down the middle of this valley in a peaceable enough fashion; all along our way there had been no rapids to speak of, and no waterfalls more than a foot or two high. Like the nar-rowness of the canyon upstream and the sheerness of its walls, the ab-sence of waterfalls here was a result of the river's having cut down through relatively homogeneous rock. If hard layers had been inter-leaved with softer ones, erosion would have slowed down at the hard strata while softer ones above would have had time to crumble away and form the ledged or terraced effect so often found along other rivers in the Canyon Lands. And the river itself, too, once it had finally ground its way through the hard rock, would have sliced with relative rapidity through the softer stuff underneath. This would have created abrupt drops—waterfalls—at the leading edge of each hard stratum.

The Paria's narrow canyon had now turned into a valley because the

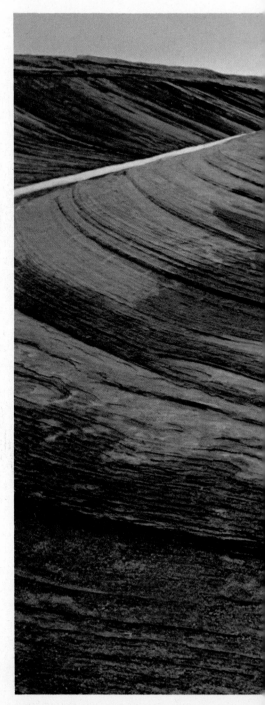

The curved contours of a petrified sand dune, looking like the markings on a giant clamshell, indicate layers of sand compressed millions of years ago by overlying rock. The dune is cemented together by ground-water minerals.

rocks through which the river flowed were no longer sandstone but an even softer shale. The occurrence of this rock formation, called Moenkopi shale, allowed the pioneers to establish a crossing of the Colorado River near the mouth of the Paria. Glen Canyon and Marble Canyon, the canyons of the Colorado that lie respectively upstream and downstream from that point, are so steep-walled as to have formed virtually impassable barriers for the early settlers. But eons ago a major fold in the earth's crust, known as the Echo Cliffs monocline, flexed the rock strata downward and relocated the softer Moenkopi shales at river level in the area where the Paria meets the Colorado. These soft reddish shales had been laid down some 200 million years ago, probably as intertidal mudflats, and then covered by other sediments. Once laid bare by the long crack in the earth's crust, the Moenkopi shales eroded rapidly to form broad slopes at the base of the Echo Cliffs. The slopes were manageable enough so that the early settlers were able to get their wagons down from the rim to the banks of the Colorado, where they could be ferried across on rafts.

Hal, who had gone on a little ahead of me, was out of sight on the other side of a small hillock when I heard him shout, "Rattlesnake!" Once on top of the rise, I could hear the steady buzz for myself. "I was following his track for a while," Hal said when I had caught up, "but then I lost it and only picked him up again when he struck out at me as I passed. Luckily I was nowhere in his neighborhood or it sure would have spoiled my day."

The rattler, a relative of the western diamondback, was yellow against the brick red color of the ledge under which it sheltered. The snake looked to be a little better than two feet long when I fished it out carefully with my walking staff to have a closer look. The thick body was tense with ridges of muscle. Once in the open, the snake struck repeatedly at my stick's hand grip, which was made of rubber soft enough so that the rattler did not injure its fangs. If it had, however, there would have been no permanent damage; rattlesnakes have reserve fangs in various stages of development folded away behind those in current use. The snake seemed to have plenty of reserve venom, too. Even on the fourth strike, milky drops trickled down the black rubber handle. This was unnerving, perhaps, but not dangerous; the venom, often fatal when injected into the body, is harmless to touch.

The striped whip snake of the evening before had been a swift, fluid animal—like a long and graceful basketball player, if you will, eeling

up through his opponents to dunk one in. But the rattler was tough, heavy and hard in the body like some slope-shouldered boxer, ready to answer any familiarity with a hook to the belly. We put the rattler back under its rock ledge and left it in command, the tense and touchy heavyweight king of a dry wash.

Only a few hundred yards farther on I came upon the tracks of another rattler. These tracks are unmistakable. Most snakes leave a fluid zigzag behind, like the track of a skier doing linked parallel turns in fresh snow. But the rattlesnake, instead of swimming along the ground by pressing its sides against small projections, drags itself forward by a rhythmic rippling of the broad, lateral scales of the belly. Thus its trail tends to be fairly straight, like that left by dragging a length of rope along the ground.

I was surprised to see signs of a second rattler so close to the first. It is rare along the river courses to see a rattler at all. I keep an eye out for them, and even go poking around in likely places, and my rough estimate is that I have come across one rattlesnake per 100 miles of hiking. In part this is because the snakes are usually nocturnal, riding out the heat and aridity of the day in underground dens. Sometimes they will lie up under the shade of ledges during the day, but mostly they prowl at night for their prey.

In this they are guided by two extraordinary organs—heat-sensing pits on either side of the head behind the nostrils. Our own largest organ, the skin, can sense the location of a nearby bonfire, but with no great precision. The rattler's two pits, on the other hand, enable the reptile to perceive in total darkness the heat radiated by a mouse six inches away. At six-inch range, a mouse produces a rise in the air temperature of $3/1000°C$. The rattler's two infrared "eyes" are thus the most sensitive temperature receptors known in the animal world. Moreover, by plotting the slightly different angles at which heat of differing intensity strikes the pits on each side of its head, the snake can pinpoint the location of an animal against a background of differing temperature. So effective is the system that a rattler deprived of the use of eyes, nose and tongue can still strike accurately at warm-blooded prey. But the snake I was tracking hadn't had a chance to bring its search-and-destroy mechanism into play; until I lost the tracks on a hard patch, they led straight from one clump of vegetation to the next, with never a detour to indicate that the rattler had flushed anything.

I set off a little ahead of Hal, so that we could both go at our own pace and think our own thoughts for a while, undistracted by another human presence. The going was smooth and easy, and the scenery was unvarying. The sun was burning down so hard by now that the dry heat sucked the moisture out of me before it could become sweat. I quickened pace and after a time shifted over into that state of semi-trance where the rhythm of walking takes over mind and body. Both elements operated efficiently enough on automatic pilot to carry out the job at hand, which was to transport a heavy pack down a river valley in the heat. From time to time the mechanism pulled up for pit stops, to drink or to wet its hat in the river. But mostly it stuck to business for the last eight miles, pausing only to register with mild irritation the growing number of signs of man's presence: here a ruined homestead invaded by a sand dune, there cows, running ahead of me in mindless fear down the valley. An abandoned cowboy's bunkhouse stood beside the ruins of a riverbank corral. In front of the empty bunkhouse a metal folding chair, all rusty, thrust its front legs into the Paria River. On the bank behind the chair, not rusting and never to rust, lay two shiny aluminum beer cans.

They made me think the conflicting thoughts I often have on coming out of the wilderness. On the one hand, it was certainly unpleasant, and even disgusting, to return again to the litter and glitter and trash of my own species. On the other hand, it had to be admitted that the sun was very hot indeed and that not too far down the trail was Lee's Ferry. And at Lee's Ferry was civilization, which I have always defined as that state of economic and social development that renders possible the existence of cold beer. Which comes in cans. And so there you are.

The Power of a River

Under controlled circumstances, the power generated by water moving downhill is easy enough to calculate. In order to estimate the hydraulic energy that might be tapped for a particular power project—such as lighting a town or operating a mill —an engineer need only determine the volume of water on hand and the grade on which it flows. But a wild river is its own unthinking engineer —with power that is most easily measurable in its physiographic consequences: the scouring of valleys and gorges, the carving of cliffs, the moving of mountains.

Few American rivers command such an awesome measure of power as does the Colorado. This rolling brown stream begins to gather its brutal force from rivulets up in the Rockies. By the time the waters reach the Canyon Lands, they have picked up millions of tons of grit and rock that act as abrasives to help carve the character of the country.

The river has been engaged in such work for perhaps 30 million years. Originally, the river now called the Colorado was an indolent stream meandering on top of a gently sloping plain. But when the Rockies and the other mighty Western ranges first reared up in the great up-heaval that reshaped the land some five million years ago, the plain rose with them, rivers and all, to a total height of about 10,000 feet.

Accelerated enormously by the increased vertical distance it now had to travel to arrive at sea level, the debris-laden water began to slice far down into the earth. Through much of its length the deepening entrenchment followed the serpentine meanders of the more leisurely ancestral Colorado, as in The Loop (pages 146-147), where in some places the waters move slowly enough to deposit gravel banks and sandbars.

But such tranquillity is deceptive, for the Colorado's flow is laced with cataracts where the river booms ahead, slashing and grinding out its chosen courses. As the river chisels downward, earth and rock break off from the canyon walls, widening or steepening the chasm. Often this debris strangles the channel, creating rapids (pages 149-151). In such angry stretches, where the waters collide with the rock, the power of the river is most dramatic—flinging white spume back toward the sky as the flow struggles to overcome these freshly deposited obstacles and get on with the task of moving its burden down to the sea.

A sloping cliff and a gravel bank (foreground), set off by an outcropping at a bend in the Colorado, attest to the river's power to move the land. The cliff was formed by the flow at the river's edge undercutting the rock. The gravel bank was deposited by detritus-bearing floodwaters; it contains enough moisture to nourish a single spray of wild rye whose roots are already working down to loosen the underlying pebbles, making them easier for the river to sweep away again.

The River's Working Partners

In the scant 10 million years of their short geological life, the Colorado and its tributaries already have cut through thousands of feet of ancient sedimentary rock to reach their present elevation. But the debris-laden waters have not performed so gargantuan a task alone. Rain and ground water penetrate into faults and fractures in the cliffs. Seeps cause the rock to dissolve and crumble, undermining the land from within. In freezing weather the water turns to ice, expands and splits the rocks, much as freezing water will burst pipes in an unheated house.

As the rocks disintegrate, winds whip away the smaller particles and rain and melting snow wash them down the slopes; tugged by gravity, the remaining debris begins a steady slide down the cliffs that have been cut by the river, eventually coming to rest on its bed. Meanwhile, the river undercuts the base of the cliff, continuously slicing, grinding and washing off material from the lowest rock layers. In this process, such softer rocks as shale wear away first; harder strata of sandstone and limestone hold out longer against the river and its erosive co-workers, giving the canyon walls their characteristic slopes and stairsteps.

Tiers of sandstone protrude from a Colorado bank; softer deposits of shale that once lay between the sandstone layers have been both weathered by rain and frost and eroded by the river.

The Uncertain Tenure of a Sandbar

The abundant mixture of silt, sand and gravel that acts like a rasp where the river runs swiftly becomes, along quieter stretches, the stuff of sandbars. These form most frequently in the backwaters, on the inside banks of broad curves. In such places the current slows down substantially, allowing gravity to overcome the momentum of the river and causing suspended debris to drop to the bottom. The deposits eventually build up until the newly created feature rises above the surface. As in an iceberg, the sunken mass far outweighs the visible part.

But the sandbars are often short-lived. Usually built at the end of the spring floods when the swollen stream subsides, the bars may begin to crumble away as early as the following autumn, especially if a dry summer has robbed the sand of its cohesive moisture. More often, they break up the following spring when floodwaters come to wash and suck away the sand. An occasional sandbar, however, can reinforce its foundations with the roots of vegetation (overleaf), and may then survive for a century or more.

Ripples carved by wind and water mottle a sandbar on an inner curve of the Colorado near Sheep Bottom. Formed after the spring floods, such a bar may remain less than a year.

A sun-dried sandbar, marked by tracks of mule deer and raccoons, breaks off in chunks along the river's edge.

Crowded tamarisks and sand willows anchor the soil of a century-old sandbar in The Loop. The durability of this bar has been ensured by its sheltered position at a point where the river makes a hairpin turn and aims its erosive power at the opposite bank.

Rapids in the Making

The rapids that constantly form in the Colorado are the result of natural damming of the river's course. Undermined by the river itself, then pried loose by frost and tumbled by gravity, a cathedral-sized rock occasionally breaks off along fault lines or fractures, toppling into the water and partly impounding it. Rapids originate at the point where the river rushes around the obstruction.

In other cases a similar dam is formed by the steady creep of boulders into the river from a talus slope *(right).* Sometimes, too, a silt-laden tributary creates an alluvial fan as it flows into the Colorado, constructing a broad underwater delta that raises the bed of the river at that point and impedes its course.

As the Colorado rushes around and over such blockages, the water pulverizes the rock and carries it away as silt, gathering abrasive power to be expended downstream. Yet the rapids themselves may endure, their fury even growing as fresh debris adds to the obstacle—steepening the grade and simultaneously increasing the volume of water held behind the natural dam.

Cracked sandstone blocks gradually breaking off the face of this cliff will, in time, drop into the riverbed and accumulate to form new rapids.

Loose debris tumbling down from the talus slope at rear constantly contributes to the build-up of these rapids in Utah's Cataract Canyon.

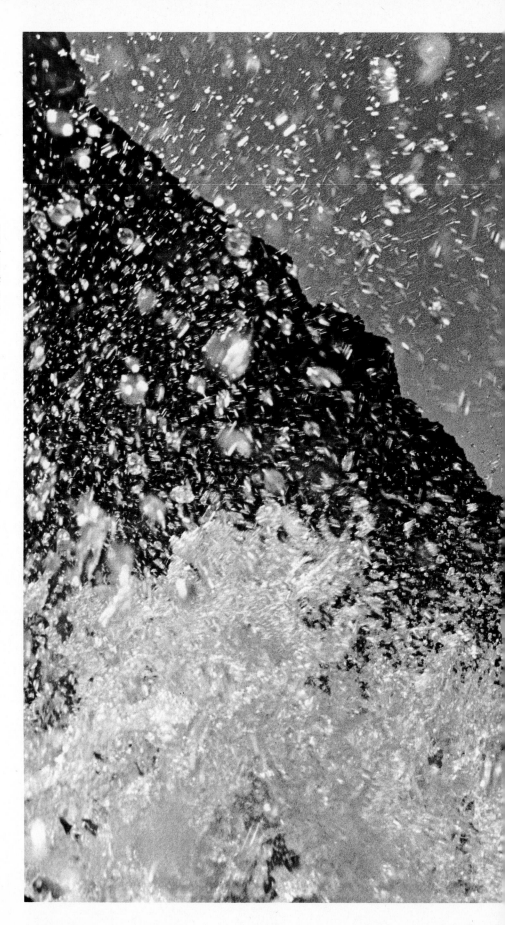

A welter of white spray spews up in the rapids at the head of Cataract Canyon. Here the water's fury is released as the riverbed makes an abrupt drop, descending 20 feet over the next mile—more than 10 times the gradient in its quiet stretches.

6/ The Sacred Mountain

---

*Far above, yet overhanging, were great yellow
crags and peaks, and between these, still higher, showed
the pine-fringed slope of Navajo Mountain with snow
in the sheltered places, and glistening streams, like
silver threads, running down.* ZANE GREY/ THE RAINBOW TRAIL

I first saw Navajo Mountain from a mesa top some 60 miles distant.
The mountain rose out of the western horizon just south of the confluence of the San Juan and Colorado rivers, barely visible in the blue
haze that blurs the distance in any canyon-country landscape. The peak
stood out all alone, a symmetrical eminence rising some two miles
above sea level.

Right away, I wanted to climb Navajo Mountain—perhaps because it
seemed singular, while the other mountains of the Canyon Lands were
plural. I could go climbing in the La Sal or Henry or Abajo mountains;
but I could simply climb Navajo Mountain. And if I did, it would become mine, all by itself, as no range of mountains ever could.

This discrete quality may have recommended Navajo Mountain to
the Indians, too. The origins and even the substance of many Navajo
myths are lost today or have survived in confusing form, with different
versions told by different clans. But in most versions of the tribe's
story of Creation, Navajo Mountain plays an important part. Long ago,
says one tale, the tribe's ancestors escaped a flood in the world beneath
this one by climbing to safety through a giant hollow reed. A mysterious old man had planted the reed in soil collected from Navajo
Mountain and six other mountains—all seven now considered sacred
by the Navajo. The seven sacred mountains were built of soil brought

from the lower world by four of the central figures in the Navajo Creation legend: Black Body and Blue Body, and First Man and First Woman. As a practical matter, this account is close enough to the real mark: certainly the masses of igneous rock that lie inside Navajo Mountain and many of the other mountains in the canyon country came from the lower world.

The Navajo further believe that the great assemblages of mountains running roughly north and south through their ancient lands represent male and female human figures. Each figure contains sacred spots where no white man must venture—in some cases, no Indian either. The eastern figure, the male one, slants southeast, head down, across the Arizona-New Mexico border. It comprises the Chuska, Lukachakai and Carrizo mountains. The huge female figure, mostly lying in western Arizona, was called the Pollen Range—corn pollen was, for the Navajo, a symbol of fertility, whose essence is female. The body of this range is formed by Black Mountain and the Hopi mesas. The head, partly in Arizona and mostly just over the line in Utah, is an isolated peak whose sacred name is Pollen Mountain. On the white man's maps, it is called Navajo Mountain.

Still a third name is Naatsis'aan, Navajo for "Hiding Place from the Enemies." In 1863, Colonel Christopher "Kit" Carson set out with a force of 736 officers and men to round up the 10,000 or so Navajo, whose presence was deemed threatening or, at least, inconvenient to encroaching whites. By using tribes traditionally hostile to the Navajo —Ute, Hopi and Zuñi—as scouts, by offering bounties on Navajo livestock, by destroying Navajo crops and orchards, Carson did the job in six months, and put an effective end to the tribe as a powerful and autonomous people. Most of the surviving Navajo, forced to surrender, were resettled in southeastern New Mexico at Fort Sumner; a few escaped Carson's net and hid safely from their enemies in the wild, isolated ravines running down from Navajo Mountain.

Unlike the other principal mountains in the Canyon Lands whose jagged profiles have been laid bare through erosion of their overlying layers of sedimentary rock, Navajo Mountain still has its sedimentary covering, stretched intact over a core of solidified lava. This accounts for the smooth, regular appearance that impressed me when I first saw Navajo Mountain in the distance. What struck me upon seeing it up close was how deceiving first appearances can be.

I got this revealing close-up when I went to climb the sacred mountain in the heat of August, five months after my first glimpse of it. Four

The scarlet bursts of a skyrocket-gilia plant, its center stalk nibbled away by grazing deer, shoot outward from a rubble of stone.

friends started the journey with me: psychologist Frank Tikalsky, botanist Hal Mackay, geologist Jim Fassett and photographer Wolf von dem Bussche. Our approach was a dirt road that led more than 50 miles through the Navajo reservation. Rabbitbrush was in bright flower among the sage, so that we seemed to be driving along in a sea of bluegreen and yellow.

The mountain lay in view for most of the way, at first a gray-blue smudge and then a more definable shape that grew larger and larger. From far away, the flanks looked as rounded and gentle as I remembered. As we drew nearer we began to see stony buttresses all around —long ridges running up from the foot of the mountain. These buttresses help give Navajo Mountain its distant illusion of gentle silhouette, but up closer they became razor-backed ridges of bare rock flaring out from the main body of the mountain, too treacherous to climb. To begin the ascent we would have to go up the inner walls of the mountain, which lay between the giant rays of stone. And these walls were precipitous talus slopes a half mile high. Hal, Frank and Wolf planned to skirt around the foot of the mountain, ending 14 miles away at Rainbow Bridge. Jim Fassett and I were headed ultimately for the same place. But we would get there by going over the mountain instead of around it—and our way led straight up one of those talus slopes.

Calling it a precipitous talus slope half a mile high may not quite get across the magnitude of the thing. Think of a wall of boulders of various sizes—some as large as a car or even a small house. The boulders lie jumbled at an angle just shallow enough to avoid landslides. Climb up this steep ramp, around and over these boulders, until you reach the height of the top of the Empire State Building: you are now only halfway up the talus slope. And when we had reached the top of the giant stone pile—arriving at a ledge some 2,600 feet up from the foot of the mountain—we were only about halfway to the real summit. "You want to know why I climb mountains?" Jim asked when I caught up to him at the ledge. He had his boots off and was massaging his feet. "I climb mountains because it feels so good when you stop."

A cool breeze blew, and we were in the shadow of ponderosa pines. Off between the trees something moved: a mare and her colt summering in the mountain. A layer of springy pine needles covered the soil. Indian paintbrush bloomed red and yellow. Here and there spires of green gentian grew waist high, the single flower stalks bent slightly uphill by the winds. Even here, well below the summit, the winds were beginning to tailor the shapes of the plants, both large and small. In the Sierra Nevada of California, naturalist John Muir once found a ponderosa pine that stood 220 feet high and was eight feet in diameter. But on Navajo Mountain the ponderosas must stop growing at the height prescribed by the winds—about 120 feet at most—or have their tops snapped off.

Most of the pines had taken the first course, but one, near the edge of the mountain shelf, had continued to grow. Some long-past storm had torn off its top and now the tree was gray and dead. The wind had broken off the remaining upper limbs, one by one, and they had been caught in the heavier lower branches. The broken limbs hung there, loosely jumbled, like a rough nest made by some giant bird.

Nearby a juniper had once stood. No doubt it had died long ago, but had remained upright for scores or even hundreds of years, as junipers do. Finally its inner strength must have given out all at once; the low tree's tortured limbs now lay around the decayed trunk like silver bones, each in the position where it had fallen. The effect was that of an exploded diagram of a juniper tree, drawn in silvery gray.

More than cottonwood, more than piñon pine, juniper is for me the tree of the Canyon Lands. A juniper with a trunk only 6 to 10 inches through may be as old as 250 years, and some are far older. Near the Grand Canyon a few junipers with three-foot trunks are said to predate Christ. All through the canyon country, dead junipers stand naked and gray, their layers of bark long since weathered away. While the tree still lives, the reddish-brown inner bark comes loose in long strips of soft, fibrous material, which the Indians taught the white settlers to use as roof thatching. The Indians also used the inner bark as swaddling for their babies and they ground the blue-green juniper berry to make it into dried cakes. The white man, further advanced along the road to something or other, puts paper diapers on his babies and uses the juniper berry to flavor his gin.

A dirt Jeep track runs to the top of Navajo Mountain, to permit maintenance men access to radio transmitting towers on the summit. It is not a public road and the Navajo do not particularly want it to become one; despite the towers, to the tribesmen the mountain still retains its sacred character. At the elevation we had reached, the track ran through the pines along a rock bench. The only spring on the mountain, if our map was to be believed, lay just south of the Jeep trail. Consequently we worked our way down a likely gully south of the trail. The little draw was full of still-white gooseberries (later they would be red, and

The scales of piñon-pine cones are flared open in early winter, indicating that they are releasing seeds—a bounty of nutritious nuts that provide food for birds and animals. The nuts were a favorite delicacy in the diet of Navajo, Apache and other Indians.

then black) and wild-rose hips, already orange and red. All this was very pretty, but nowhere near as pretty as a spring would have been. When we finally did find the spring, just at dark, it proved to be north instead of south of the trail.

The water running to the surface was cool and pure. Jim, a connoisseur of desert water, recognized it as of late Jurassic vintage—from the Bluff-Summerville mudflats laid down in windblown dunes under our feet some 150 million years ago. Water from different rock formations, it seems, sometimes has different tastes, depending on the cementing material between the grains of sand—or on the kinds of minerals that have been deposited in the rock by ground water over the ages. Some of the water is so sulfurous that a man has to be very thirsty indeed to get the stuff down.

Along the corridors of pines, we could still make out the pale forms of the gray mare and her colt. Some small animal scuttled near us, but out of sight—probably a squirrel or a chipmunk. Both these small rodents act unwittingly as the nurserymen of the pine forests by gathering pine cones and burying them to eat later. Often enough, the animals then forget about the cache or get killed by some predator. Seeds thus planted often come up better than those scattered normally on the ground. The hoards are at just the right depth for germination, and squirrels and chipmunks try to select perfect cones for their larders. Furthermore, if a forest fire should come, the caches are safe underground, ready to reseed the forest after the burn.

By the time we had had our drink, the day's heat was broken and the colors of the woods in the dusk were soft green and brown. Scattered around the base of each ponderosa lay the curious platelets that scale from the tree's bark. They are reddish brown and just about the size, thickness and shape of pieces from a jigsaw puzzle. The breeze that often accompanies nightfall in the canyon country was rising. We had made camp and lit a fire; water for supper was boiling and there was plenty more water in the spring for tomorrow. Jim had been right; it did feel good when you stopped.

Next morning, we started up the trail that led the rest of the way to the top of the mountain. Now and then we paused at one of the switchbacks to look at the view spread below us. Through the pines the sun glared white on the sterile rocks that ran on to the horizon. During the ascent, the cathedral effect of the ponderosa forest below had slowly given way to a cluttered, desperate look. Now fewer ponderosa remained and they were much smaller. The forest floor was thickly set

The smoothly rounded contours of Navajo Mountain in Utah rise above the grasses and sandstone hummocks of the high plains of Arizona. While the extraordinary clarity of the canyon country's dry air makes the mountain seem nearby, it is actually 13 miles away—and eight miles distant from the twin tabletops of Cummings Mesa silhouetted on the horizon at left.

with ground juniper, young aspen and white fir. The Jeep trail we were following led us eventually to the two transmitting towers on the highest part of the mountain. The wind sang in the guy wires of the towers. No one stays up permanently to watch over them, and all the outbuildings were shut up tight—except for a two-holer of plywood painted white. In its ceiling was a heat lamp, powered by the transmitters' generators, which filled the interior with an orange-red, warm glow.

As we had approached the mountain, the towers had been too insignificant to be seen—a couple of bristles upon its great back. And once we were on the mountain, the slope and the forest hid the towers from us. We knew the towers were there, but it had still been a jarring thing to come upon them suddenly in the middle of the wilderness: girders thrusting up in unnatural straight lines from the gnarled forest; small and homely, utilitarian outbuildings; the hum of the generators. But we walked on a little, and in five minutes the radio towers had disappeared from our universe.

Even during August, in the shelter of the woods, a chilly wind cut and pushed at us; when the winter storms came to the mountaintop, the weather would be murderously harsh. There was no need to guess at this. The forest around us gave evidence enough. Aspen are normally tall trees whose slender, white shafts rise as naked as palm trunks before they finally put out a crown of branches high overhead. But here the winds allowed the aspen to grow to no higher than 15 feet or so. Instead of being slim and stately, those trees in exposed positions were so thick-bodied that their few twisted branches seemed absurdly small for such squat and massive trunks. In one little clearing stood an aspen that, in its sapling days, had evidently been forced into a corkscrew shape by the weight of snow on its crown. As it had grown, the sapling trunk held the strange shape: where it came out of the ground the mature tree still curled like a pig's tail—before going on about its vertical business.

Navajo Mountain is fairly flat on top, so that we had easy going until we reached the northwest edge. Here the trees gave way at last and we could see the view. Lake Powell lay before us, sending blue fingers into canyons on both sides. The largest of these ran up the San Juan River, whose confluence with the Colorado disappeared under the water 10 years ago, after Glen Canyon Dam began to fill in 1963.

Another of the lake's fingers went up Forbidding Canyon, which is, in consequence of the dam, no longer particularly forbidding. Powerboats can now go up the canyon and along one of its tributaries nearly to Rainbow Bridge—formerly accessible only to those willing to hike or ride horseback over the 14 miles of rough trail our companions were taking around the base of Navajo Mountain. With our field glasses, we could see the bridge, the world's largest natural arch. But it was barely visible. You had to know just where to look, and then it seemed to be not much more than an insignificant knothole in the red rock, dwarfed by the scope and sweep of the land.

On the mountaintop where we stood, frost had worked to erode the surface of the mountain in unique and fascinating patterns. The freezing and thawing had split the skin of hard sandstone—a dense rock composed of quartz grains—into millions and millions of chunks. The rocks were not large—most of them were no bigger than a basketball—and thus, in mass, had a certain fluidity that larger boulders would have lacked. Here and there along this northwestern face of the mountain, sheets of the gray sandstone curved over the edge like frozen rivers and plunged downward.

These streams of rock had separated to avoid the obstacles formed by larger boulders, had come together again once past them, had broken up here and there into smaller rivulets, and had behaved in general just like flowing water. And for good reason. In the immense framework of geological time these might seem to be rivers—raging torrents of rock boiling down a steep mountainside. If we have trouble seeing the matter in this active way, it is because we are such ephemeral creatures, here one century and gone the next. A being with a life span of 1/1,000th of a second might have just as much trouble comprehending the true nature of Niagara Falls. To him, the falls would be a picturesque arrangement of sheets and droplets of water with the appearance of hanging static in the air.

There on the brink of the rock cascade, colors seemed more distinct somehow and the outlines of objects more sharply defined. Forms stood out from one another, and textures could almost be felt with the eyes. The trunk of a weather-beaten aspen was smooth, nubbly, almost waxy. A large fallen spruce, dry as paper, felt the same warm, brown, shaggy way it looked. Underfoot, the sandstone rocks rang with a bright sound quite unlike the dead, heavy clunks of shale and limestone.

The wind leaned against us with a steady push and the sky was blue for a hundred miles. We left the wind above and behind us as we started down and across one of the rock rivers. When we reached the precipitous slope on the far side of it, we were shaded by spruce and fir. Wherever we could, we tried to keep off the unsteady rocks and in the trees, so we would have handholds. The descent went on and on, as steep as an expert ski run, a mile of vertical drop in two and a half horizontal miles. At length we emerged onto a gentler slope, thickly set with juniper, and finally came down off the mountain into the upper reaches of Oak Canyon.

Here we encountered a stream fed by water from the heights and supporting tall reeds, cattails, redbud and cottonwood. The growth was so thick and tangled that it forced us out of the canyon after several miles, and up onto the flats above. There we would have intersected the trail that led to Rainbow Bridge, but we were caught by approaching darkness before we could reach the junction, and made camp in Rainbow Bridge Canyon.

Little spadefoot toads came out in the dusk and were everywhere on the ground. Less than an inch long, they must have been tadpoles in the small nearby pool not long before. They had, therefore, already lived through a good deal of dark and bloody history. Their life cycle begins when rain or floods soak the baked mud in which adult spadefoots bury themselves to last out the dry season. Using the spade-shaped rear feet for which they are named, the adults dig themselves free.

The first male toad to reach water jumps into it and begins to croak, a sound that sympathetic observers have likened to a greatly amplified recording of somebody running his fingernail along the teeth of a comb. Although studies are so far lacking on just what this noise might signify to a comb, to a toad it is a love song. Each toad that hears the call, which carries some two miles, starts to hop toward the source. As other males arrive at the water, they add their voices to the din. By now toads of both sexes are floating around in the water together, the females maintaining a discreet silence. It is dark, and expectancy runs high. Male toads grab any toad they may bump into in the dark. If the grabbed toad is another male, he signals the fact by croaking and the grabber lets him go. If the grabbed toad does not croak, mating proceeds.

By daybreak, the females have already laid their eggs, which hatch in two days. This is much faster than is normal for toad and frog eggs; the whole reproductive cycle of the spadefoot toad is greatly speeded up to take advantage of water that might vanish quickly. One day after hatching, the tadpoles start to feed—some of them doubling their weight each day for the first two days. In each little pond there may be thousands of tadpoles; a puddle not two feet across may hold hun-

A tough juniper, contorted and stunted after decades of subfreezing winters and the 100° heat of summer, clings to its bare root bed on a mesa top. In fall and winter the juniper produces fat berries (above). Ute and Paiute Indians once ground mush and cakes from the fruit, and boiled a fragrant wax out of the berries' coating.

dreds. The stronger tadpoles attack this classic case of overpopulation head on by eating the surplus. "The growth and development of living tads," one herpetologist has written, "is greatly stimulated by a diet of dead tads when the water level in the pond is falling dangerously low."

This, then, was the harsh but instructive story of the tiny toads hopping around our camp. I tried to explain all of it to Jim, but noticed after a time that he had unaccountably fallen asleep just as I arrived at the most interesting part of my thesis, which deals with the rising population and falling water table of such desert cities as Phoenix, Albuquerque and Tucson.

Next morning we moved through the canyons toward Rainbow Bridge, where we were to meet our friends who had hiked around the mountain. Along the way we came on a Navajo hogan built in the old style, of driftwood retrieved from a canyon stream and topped with juniper-bark thatching. These were the canyons, protected by the mountain on one side and the Colorado River on the other, that had afforded a safe hiding place for Navajo escaping from Kit Carson's roundup of their tribe more than a century before. The old hogan might have been there that long ago; it could have been built even earlier. A century is no time at all, in the dryness of the canyon country, for a wooden structure to endure.

We found our friends at a place called Echo Camp, just short of Rainbow Bridge. The camp had once accommodated tourists who traveled to the bridge in packtrains, taking off from a trading post 14 miles away. Now Echo Camp lies deserted, made obsolete by Lake Powell and the boat-docking facilities just a few easy yards from Rainbow Bridge. All that remains of the camp are a few sheds, a fireplace, a corral and a great many iron bedsteads standing around in the open. These artifacts are grouped around a shallow, diminished pond fed by a trickle of water from the canyon walls that climb overhead.

While I waited for my canteen to fill at one of the trickles, my eye was caught by a black hair bending and contorting beneath the surface of the pool. I would have dismissed it as a rootlet in the current if the pool had not been dead still. But the hair was alive—one of the bizarre creatures known variously as Gordian worms or horsehair snakes. Even upon close inspection, these animals, a species of Nematomorpha, look precisely like a hair from a horse's mane or tail.

This one twisted, eyeless and directionless, in the water in front of me. Not only is the horsehair snake blind, but it has no more than a minute pore for a mouth, and sometimes even the pore is lacking. It does not feed at all, has no circulatory system and no excretory system. Most of its life is passed as a larva in the stomach of some insect, frequently a grasshopper. As a larva it absorbs nourishment through its skin. When it emerges as a blind, hairlike adult, its only business is to reproduce and die.

I caught the slowly convoluting hair in a cup and examined its eerie dance for a minute before putting it back in the pool to go about its odd business. Then I went about mine, which was to stretch out in the shade on one of the rusty bedsteads.

What had happened to Echo Camp and to the Navajo hogan, I thought, will happen to Tucson, Phoenix, Albuquerque one day, when their water finally gives out (as I would have told Jim if he had managed to stay awake). The desert does not destroy man's leavings as the jungle does. Beams and pillars don't rot in a season or two; banyan tree roots don't split the stones nor do vines crawl over ruins, prying brick from brick and blanketing the rubble from view.

The desert's way is to preserve things, as if in amber. The objects remain, but their colors are muted into those of the land itself. Paint flakes off wood and iron in the sun. The wood turns the silver of dead junipers, and the iron rusts to the color of Navajo sandstone. The signs of man's brief tenancy are not removed: they are made into a new and harmonious element of the vast land.

Back up the canyon, blue in the distance, rose the huge bulk of Navajo Mountain. "An island in the midst of a sea of waterworn and windworn, brilliantly colored sandstone," geologist Herbert E. Gregory called it half a century ago. The island and the sea, between them, still seemed adequate to absorb whatever flyspecks man might leave.

Spirits of the Rainbow Bridge Trail

In southern Utah, not far from the Arizona border, photographer Wolf von dem Bussche walked out across the red stone desert and through the canyons from Navajo Mountain to Rainbow Bridge (pages 178-179). A stranger in the land, he came away with the feeling that he had been in the presence of the supernatural—a feeling that would have seemed quite proper to the region's Indians, who say they have been living here ever since their legendary ancestors emerged from beneath the earth. To the Navajo every butte, every canyon, cliff and cloud has its own life and meaning; the landscape is filled with presences that leave no record on film but exist nonetheless.

In the desert country, the aura of a supernatural presence does not present itself in drifting shapes of the fog or sounds in the wind. It is borne upon silence and appears in clarity of air, in oceans of atmosphere so pure and still that objects known to be a day's journey away seem far closer. In this transparency any man has the chance to see the world—and himself—as though for the first time.

On the trail von dem Bussche followed to Rainbow Bridge are junipers that have been rooted in the rocks for hundreds of years; and there are dead ones even more ancient that will not fall, but stand with arms still raised in greeting or farewell. The silent sun pounds the walls of the canyons with the power of a sledge. Overhead sail little clouds that rub their shadows across the stone. As they pass, the rock momentarily darkens, then brightens and resumes sending up its waves of shimmering heat. The colors of this world are in constant change, purple to vermilion, ocher to gold, and everywhere can be sensed invisible forces as old as the Creation.

Sometimes, according to the Navajo belief, these forces or presences make themselves felt directly on the mystical landscape; the enormous arch of the bridge, 278 feet wide and 309 high, was not cut in the sandstone by the meanders of streams during a million years or more. Rather, according to one Navajo account, it was created in a brief moment by a spirit. He heard the prayer of another supernatural being who had been trapped in a canyon by a sudden flood, and in response threw down a rainbow on which the kindred spirit reached safety. Beneath the traveler's feet the rainbow turned to stone. And so it remains.

RABBITBRUSH BELOW NAVAJO MOUNTAIN

CUMMINGS MESA, FAR OFF

JUNIPER IN A CLEFT ROCK

A STONE PINNACLE FALLING INTO RUIN

DEAD TIMBER AT THE EDGE OF A CANYON

CARVED SANDSTONE IN CLIFF CANYON

A TRACERY OF JUNIPERS AGAINST CLIFF CANYON WALL

A SWIRL OF CLOUDS ABOVE A BUTTRESS OF NAVAJO MOUNTAIN

AN ARRAY OF EARTH COLORS NEAR NAVAJO MOUNTAIN

RAINBOW BRIDGE IN EARLY MORNING